W9-BVX-019

Macachiavellian Intelligence

MACACHIAVELLIAN INTELLIGENCE

How Rhesus Macaques and Humans Have Conquered the World

DARIO MAESTRIPIERI

The University of Chicago Press
CHICAGO AND LONDON

Dario Maestripieri is associate professor of Comparative Human Development and Evolutionary Biology at the University of Chicago. He has studied the behavior of rhesus macaques and other primates for over two decades at the University of Roma La Sapienza, the University of Cambridge, the Yerkes National Primate Research Center of Emory University, and the Caribbean Primate Research Center of the University of Puerto Rico. He is a Fellow of the American Association for the Advancement of Science and has received awards from the Accademia Nazionale dei Lincei, the American Psychological Association, and the National Institute of Mental Health. He is the author of over 125 scientific publications and editor of the book *Primate Psychology* (Harvard University Press, 2003). His research on macaque and human behavior is widely cited in the popular media.

The University of Chicago Press, Chicago 60637
The University of Chicago Press, Ltd., London

Printed in the United States of America

16 15 14 13 12 11 10 09 08 07   1 2 3 4 5

ISBN-13: 978-0-226-50117-8 (cloth)
ISBN-10: 0-226-50117-5 (cloth)

LIBRARY OF CONGRESS CATALOGING-IN-PUBLICATION DATA

Maestripieri, Dario.
Macachiavellian intelligence : how rhesus macaques and humans have conquered the world / Dario Maestripieri.
p. cm.
Includes bibliographical references and index.
ISBN-13: 978-0-226-50117-8 (cloth : alk. paper)
ISBN-10: 0-226-50117-5 (cloth : alk. paper)
1. Rhesus monkey—Behavior. 2. Rhesus monkey—Psychology. 3. Machiavellianism (Psychology) 4. Social behavior in animals. 5. Psychology, Comparative. I. Title.
QL737.P93M34 2007
599.8'64315—dc22

2007012023

♾ The paper used in this publication meets the minimum requirements of the American National Standard for Information Sciences—Permanence of Paper for Printed Library Materials, ANSI Z39.48-1992.

Contents

Chapter 1
THE SECRET OF OUR SUCCESS

The Most Successful Primates

The U.S. Census Bureau estimates that by the year 2050, the world's human population will have quadrupled in size since 1950. You can find people almost anywhere on the globe and in all kinds of habitats, from the Inuit of the Arctic to the !Kung of the Kalahari Desert in Africa. *Homo sapiens* has clearly been the most successful of the more than three hundred primate species currently living on our planet, and it's no secret that our big brains and sagacity helped facilitate our success.

By the same criteria of population size and geographic distribution, another very successful primate on this planet is a monkey called the rhesus macaque. The rhesus macaque, however, is not one of the smartest primates. Other primates—the great apes—have bigger brains and are smarter than rhesus macaques, but unfortunately they are all on the brink of extinction. So being smart is not by itself a guarantee of success in this corner of the universe. There are different kinds of intelligence and different ways to use it.

This book is about rhesus macaques and what they have in common with people. There are more facts in this book about monkeys than about people, but the book is really more about people than monkeys. Why rhesus macaques are the way they are is an interesting question, but the fact that human beings often act like rhesus macaques is even more interesting. Those readers who already see the monkey in the mirror may discover that this monkey looks more like a rhesus macaque than they ever thought possible, but they may not like what they see. Those who are used to seeing only themselves in the mirror will like it even less. Finally, for those who have never looked in the mirror at all, this may be a riveting place to start.

The Buddy Story

An adolescent rhesus macaque male has been captured by a group of researchers and taken into a dark concrete building for testing. He's given a sedative and sleeps for a good hour on the floor of a cage. When the monkey's eyes open, he stands up and drowsily assesses his location. More time goes by, and the monkey is now alert, walking around in the cage and looking anxious to get out of there. A door is opened, and the monkey darts out and back into the enclosure where the rest of the group lives. A hundred pairs of monkey eyes look at the newcomer for a second, then look away. No reason to be alarmed; it's just Buddy. It was his turn today, and he's back. The older females go back to their grooming, the alpha male resumes his nap, the infants continue to play on the jungle gym. Buddy's favorite playmate walks up to him and seems eager to engage him. He pushes Buddy and then runs off, looking for a chase. But wait—Buddy isn't coming. He fell on his side and is slowly getting back on his feet. Something's wrong. A hundred pairs of monkey eyes look at Buddy again. A tall and muscular subadult male, a bully, walks up to Buddy and stares him down. Buddy looks at him with a puzzled expression for a couple of seconds, then turns his head away. The bully bites Buddy's arm. Buddy screams in pain and runs away. But he is slow, too slow. The bully quickly catches up with him and bites him again, this time on his ear. More screaming. Two other adolescents—Buddy's playmate is one of them—and an adult female run toward

Buddy, looking excited. He runs away, but they get him, and he's on the ground again, and they are all over him, barking and screaming, grabbing his arms and face, and biting his fingers and tail.

Everything has happened quickly, but the researchers have been watching, and the moment they see Buddy fall awkwardly they know that they have to get him out as soon as possible. They catch him and put him in a cage by himself. He looks frightened but has no injuries. Two hours later he returns to the group. His playmate and another adolescent walk up to him and grab him. He grabs them back, and the three of them wrestle. Then Buddy gets chased, but this time he runs quickly and is not caught. As he runs, he inadvertently bumps into a young infant and knocks him down. Immediately the infant's mother arrives, picks the infant up, and threatens Buddy with a stare and a wide open mouth. Buddy shows his teeth to the mother and raises his tail, exposing his genitalia to any other monkey who might be behind him. Nothing else happens. The mother turns around and walks away. Buddy walks to the food pile, grabs an apple, and starts eating. No one pays attention to him now.

Buddy has spent every day of his life in the enclosure with all the other monkeys. They all eat the same food and sleep under the same roof. Buddy's family has low social status, but there are other families below them in the hierarchy. He spends a lot of time with adolescents from other families and has been seen hanging out with older males and females as well. They were there when he was born. They held him and cuddled him when he was an infant. They have watched him grow, day by day, every day of his life. Yet, that day, if the researchers had not taken Buddy out of the group, he would have been killed. His mother and aunts would have tried to protect him, but probably to no avail.

Buddy had not fully recovered from the anesthesia when he was first reintroduced into his group. The others could immediately tell there was something wrong with him. He wasn't running as quickly as usual. He didn't respond to a threat with a submissive signal. He didn't run back to his mother seeking protection. He was weak and vulnerable. The behavior of the other monkeys changed swiftly and dramatically—from friendliness to intolerance, from play to aggression. Buddy's vulnerability became an opportunity for others to settle an old score, improve their position in the dominance hierarchy, or eliminate a potential rival for good. In rhesus macaque society, maintaining one's social status, being tolerated by others, and ultimately surviving at all may depend on how quickly one runs and how effectively one uses the right signal, with the right individual, at the right time. A rhesus macaque can wake up one morning, feel a little drowsy, and find himself in danger of being killed by his best friends.

Macachiavellian Intelligence

Imagine a society in which everybody walks around with a loaded rifle. The citizens of this society—some more than others—must constantly watch their backs and avoid any situations that may lead their comrades to fire their weapons at them. Rhesus macaque society has a strong hierarchical structure, and individuals of high status use their power against their subordinates without mercy. Genuine altruistic behavior is shown only with one's closest relatives. Social relationships with everybody else are governed by the laws of the market: you scratch my back, I'll scratch yours. If you are kind to someone, you expect something in return, typically sex or help. Social opportunism and manipulation are the rules of the game. Yet the bonds between family members are strong, and the group is cohesive and ready to fight against its enemies, whoever they might be.

Niccolò Machiavelli wrote his famous book *The Prince*[1] in 1513 to instruct Lorenzo II de' Medici, his patron and the ruler of Florence, in the art of politics. It explains how to pursue and maintain political power and exploit everyone and everything in the process. Following Machiavelli, social opportunism came to be referred to as Machiavellian intelligence. Rhesus macaques had already been using Machiavelli's recommendations in their daily lives for thousands of years.

If Macachiavellian intelligence is what people and rhesus macaques have in common, could that be one of the reasons for their success? Could it be that Macachiavellian intelligence explains why some species or societies are more successful in the competition for survival than others? Rhesus macaque society is organized and functions like an army. Armies are the type of social organization people use to conquer other groups of people, their land, and their possessions. Armies throughout the world and throughout human history have tended to have the same hierarchical structure and follow the same rules of behavior. Is that just a coincidence?

Maybe the Macachiavellism of people and rhesus macaques has nothing to do with their success. Charles Darwin once wrote, "He who understands baboon would do more towards metaphysics than Locke."[2] Without taking anything away from baboons, understanding why rhesus macaques behave the way they do may tell us something about human nature, metaphysics, and perhaps the future as well. By the time human beings start the global nuclear war that will destroy our civilization, there won't be any great apes left for Earth to become the Planet of the Apes. But chances are there will still be plenty of rhesus macaques around.

Chapter 2
THE WEED MACAQUE

The Average Monkey

If people were asked to think about a monkey, they would probably form an image in their minds that is either that of a young chimpanzee or a composite of all the monkeys and apes they've seen at the zoo or on TV. This imaginary "average" monkey would probably look a lot like a rhesus macaque. Because of their average looks, the fact that there are so many of them around, and their lack of any flashy behavioral peculiarities, such as using tools to eat termites the way chimpanzees do,[1] rhesus macaques never make the news headlines and generally don't attract the attention of nature documentary makers. The specialty of rhesus macaques is Machiavellian intelligence, but that's hard to show in the background of a TV screen behind Sir David Attenborough.[2] Anybody who saw a rhesus macaque walking down the street would probably not know what kind of primate it is, and given the stereotypes we all have about monkeys, might be tempted to offer the monkey a banana. That would be a big mistake,

FIGURE 1 Adult female rhesus macaque on Cayo Santiago. (Photo: Dario Maestripieri.)

because the rhesus macaque would readily accept the treat and then jump on the person's back and aggressively demand more.

The rhesus macaque is a medium-sized monkey. Adults are about half a meter tall and weigh between 5 and 8 kilograms. Males are about 5–10 centimeters taller and 2–3 kilograms heavier than females. Their bodies are covered with brown fur, while their faces and rumps are pink or red and have no fur. Like other monkeys, and unlike the great apes, rhesus macaques have a tail, which they often raise as a white flag to signal surrender, but which more often than not ends up being bitten by another rhesus macaque. Rhesus macaques live about twenty to thirty years. Females reach puberty at three or four years of age, after which they give birth to an infant every year or every other year. Males reach puberty about one year later than females, and whether they become fathers after that, and how often, depends on their skills and their good or bad luck.

Who Are Rhesus Macaques and Where Do They Come From?

Rhesus macaques are one of nineteen different species of macaques (or maybe eighteen or twenty, depending on who you ask).[3] The first part of their scientific name is the same for all macaque species—it's the Latin word *Macaca*, the name of their genus. Their last name is different for each species, and for rhesus macaques, it's the Latin word *mulatta*, which means dark or black. So the scientific name of the rhesus macaque is *Macaca mulatta*.

Until August 2006, most people in the United States had probably never heard the word "macaca" before. Then, on August 11, a U.S. Republican senator named George Allen, who was campaigning for his reelection in Virginia, used that word to refer to a young campaign staffer of Indian ancestry who worked for Allen's Democratic opponent. No one who heard Allen that day probably knew what the senator meant by calling someone "macaca," but after some frantic research on the Internet, reporters discovered that a macaca is a kind of monkey, and that this word is used as a racial slur against dark-skinned people in some parts of the world. Outrage at the senator quickly followed. Allen lost the Virginia Senate race, and "macaca" was named the most politically incorrect word of 2006 by the Global Language Monitor, a nonprofit group that studies language usage around the world.[4] If politicians knew more about the Machiavellian intelligence of rhesus macaques, they would probably call one another "macaca" all the time, but mean it as a compliment.

The real macaques belong to a group of primates called Old World monkeys because they are found only in Africa and Asia, in contrast to another group of primates called New World monkeys, which live in Central and South America. All the primates that once roamed free in Europe and North America have long been extinct—except humans, of course. Humans are also the only primates that made it to Australia. The Old and New World monkeys are distinguished from the prosimians—a group of primitive primates that includes lemurs and bushbabies—and from the apes. There are two kinds of "lesser apes"—gibbons and siamangs—and four species of "great apes": chimpanzees, bonobos, gorillas, and orangutans. Chimpanzees and bonobos are the primates that are most closely related to humans. They share about 98 percent of their genetic material with people, and their ancestors split from our own ancestors about 5–6 million years ago, which in evolutionary time is like the day before yesterday. Chimpanzees are so close to people that a woman recently sent a letter to the popular syndicated column "AskMarilyn," asking, "Is it true that my husband is genetically more similar to a male chimp than to me?" Marilyn's answer was along the lines of "Maybe, but don't feel bad, we've all dated guys like this."[5] The Old World monkeys share about 95 percent of their genetic material with humans, and their ancestors split from our ape ancestors about 25 million years ago. So rhesus macaques are genetically close to humans, but not as close as chimpanzees are.

The evolutionary history of macaques is a textbook case of what biologists call an adaptive radiation: a process by which organisms colonize new environments, adapt to the local conditions, and diversify into different species. Macaques originated in Africa from baboonlike monkeys at about the same time our hominid ancestors split from the other apes. The macaque ancestors migrated from northern Africa to Europe when the two continents were still well connected by land bridges. From Europe, the macaques migrated eastward and into the Asian continent. Those that were left behind in Africa gradually shrank in numbers and geographic distribution. Their descendants are now known as Barbary macaques, the only macaque species currently found on the African continent, where they occupy small areas of Algeria and Morocco. As macaques spread into Asia and colonized the entire southern part of the continent, going as far as China and Japan, they gradually became extinct in Europe because of climatic changes and competition from other primates, including Neanderthals and other hominids. So today, with the exception of Barbary macaques, all of the wild macaques live in Asia.[6]

As the macaques spread from Europe into Asia, they differentiated into many species. Taxonomists recognize four different groups of macaque species.[7] The species within each group are genetically and morphologically more similar to one another than to the species in the other groups. The rhesus macaque and its close relatives—the Japanese macaque, the Taiwanese macaque, and the long-tailed macaque—are the most recent group of macaques to have evolved and colonized Asia. Rhesus macaques, in particular, have been very successful in this colonization process. Today wild rhesus macaques can be found almost anywhere throughout mainland Asia: Afghanistan, India, Thailand, China, Pakistan, Bhutan, Burma, Nepal, Bangladesh, Laos, and Vietnam. They are also found in almost any type of habitat, including tropical forests, dry and semidesert regions, swamps, and mountains up to 4,000 meters high. Rhesus macaques have been successful at displacing other macaque species from their native forest habitats as well as colonizing new habitats where no other macaques had gone before. So today, according to Charles Southwick—a biologist who studied rhesus macaques all his life—rhesus macaques are the second most widespread primate species in Asia, and also in the world, with the top spot, of course, being taken by humans.[8] Today there are rhesus macaques in Europe and North and Central America because people brought them there and keep them there. Europeans and Americans have tried to bring home and keep in their yards almost every primate species on the planet, but for most of these species, that new habitat just didn't work out. For rhesus macaques, however, it worked just fine.

Weed Macaques

A key component of rhesus macaques' success has been their ability to adapt to changes in the environment induced by people and to people themselves. A few other macaque species have done the same, but not as successfully as rhesus macaques have. A few years ago, some primatologists proposed to call those species that have successfully adapted to human presence "weed macaques."[9] Weeds are plants that spread and thrive in places where they did not evolve, typically in areas that have been significantly altered by people's presence. The characteristics of weeds are rapid growth, frequent and successful reproduction, ability to disperse near and far, and ecological adaptability to a wide range of habitats. Weeds are competitive, persistent, and difficult to eradicate. Rhesus macaques share some of these characteristics with weeds, and also have the behavioral attributes of animal pests, such as rats—for example, an omnivorous diet, curiosity and a tendency to explore, ability to move at high speed on the ground, and a gregarious and aggressive temperament.

Many species of primates and other animals are very sensitive to even minor disruptions of their environment, and their populations have declined quickly in response to human presence and activities. The spread of agriculture and pastoralism in Asia in the last few thousand years, along with the urbanization of large areas, has pushed many animal species to the brink of extinction. In contrast, the weed macaques quickly adapted to these new habitats and now thrive in them. Some researchers have estimated that in northern India, almost half the local population of rhesus macaques lives in villages, towns, temples, and railway stations. Most of the time, they just hang out on roadsides and in close contact with people. The rhesus macaques have learned to take advantage of people and raid their crops, eat their garbage, or steal food directly from their kitchens. The macaques that live in temples have somehow figured out a way to get the local people to worship them and bring them all kinds of food, including bread, bananas, ice cream, cakes, and cans of Coca Cola.[10]

The rhesus macaque is the quintessential weed macaque. These primates can survive and reproduce almost anywhere. They don't seem to mind if people are around and doing all the things that would freak out any other wild animal. Rhesus macaques' adaptability and resourcefulness also helped them when they left their native Asian continent and, like Christopher Columbus, discovered America.

Primates of the Caribbean

In the mid-1930s, Clarence Ray Carpenter, an American biologist who had studied wild primates in Panama and Asia, had the idea of introducing Asian primates to an island in the Caribbean, where their behavior could be observed for long periods and they could be bred and used for biomedical research. Cayo Santiago, a 15.2-hectare island 1 kilometer off the southeastern coast of Puerto Rico, seemed perfect for the purpose.[11] In 1938, Carpenter set off to collect five hundred rhesus macaques from various provinces of India and twenty to thirty gibbons from other East Asian countries. The monkeys were loaded onto the USS *Coamo* in Calcutta, and after a 14,000-mile journey that took forty-seven days, they arrived in

Puerto Rico. The gibbons did not work out on Cayo Santiago, and they had to be caged and later shipped to various zoos in the United States. At that time, it was not known that gibbons live in male-female pairs and that each pair defends a territory much larger than Cayo Santiago from other gibbons. Carpenter did know that rhesus macaques lived in large groups in which the females outnumbered the males, so he was careful to bring far more females than males to Puerto Rico. Lack of knowledge about the proper number of males and females in groups of hamadryas baboons had caused massive fighting and killing among males when, years earlier, a large group of these baboons had been assembled at the London Zoo.[12]

Carpenter, however, was naïve enough to think that the five hundred rhesus macaques that had been collected from seven different provinces in India would automatically sort themselves out and form groups in relation to where they had come from. Unfortunately, the monkeys had other plans, and those plans involved killing each other. Many of the hundred or so infants that had been captured and shipped with their mothers were killed, and several adult males were driven into the sea, where they drowned. Amid

FIGURE 2 Rhesus macaques on Cayo Santiago. (Photo: Dario Maestripieri.)

FIGURE 3 Researchers collecting behavioral data on Cayo Santiago. (Photo: Dario Maestripieri.)

all the fighting and killing, there must have been some sex going on as well, because six months later the first infant was born on the island. Its arrival disproved the prediction of an expert in rhesus macaque reproduction named Carl Hartman, who had guaranteed that rhesus macaques would not breed in Puerto Rico.[13] Between 1940 and 1942, these weed macaques produced almost two hundred new infants on Cayo Santiago. Feeding the growing population became a problem. Large amounts of fruit and vegetables grown in Puerto Rico were not enough and had to be supplemented with fox chow imported from St. Louis.

During the years of World War II and until the mid-1950s, interest in the rhesus macaque population on Cayo Santiago dwindled, and many monkeys died from starvation, cannibalism, or disease, while a few brave individuals managed to swim the kilometer to the coast of Puerto Rico. Today there are rhesus macaques running around many areas of Puerto Rico, and some of them have recently been sighted at the outskirts of San Juan—about 30 miles from Cayo Santiago—making the city's 1.5 million residents very uncomfortable. In 1955, the U.S. National Institutes of Health began supporting the rhesus colony, and they have continued to do so to this day.

Every year, a number of rhesus macaques are shipped to NIH, to the University of Puerto Rico, and to many other U.S. research institutions engaged in biomedical research, which helps keep the growth of the population in check. U.S. research institutions also continue to import thousands of rhesus macaques from Asia. Until India banned the export of primates in the 1970s, most rhesus macaques were imported from that country; since then, rhesus macaques have been imported from China and Nepal.

The Biomedical Monkey

Rhesus macaques are by far the most common primates used in biomedical research. Every large research university in the United States probably has some rhesus macaques hidden somewhere in the basement of its medical school. The U.S. Army and NASA have rhesus macaques too, and for years they trained them to play computer video games to see whether the monkeys could learn to pilot planes and launch missiles. The reason rhesus macaques are so popular in research laboratories is the same reason most people tend to have the same few house plant species in their living rooms: they are very low maintenance and they grow and reproduce under almost any conditions. Whether rhesus macaques are kept indoors or outdoors, in northern Europe or in the Caribbean, they have their own internal biological clock that tells them when it's time to mate and when it's time to have babies. In the wild, rhesus macaques produce babies in the months of the year when food is most abundant, typically during the rainy season, if there is one. In research laboratories around the world, rhesus macaques continue to breed seasonally even if they eat the same food every day of the year. In Puerto Rico, rhesus macaques mate in the spring and summer and give birth in the fall and winter. In the continental United States, the seasons are inverted: mating is in the fall and winter and infants are born in the spring and summer. The reproductive biological clock of rhesus macaques is sensitive to day length and temperature, but—just like the rest of their bodies and their behaviors—it is very resilient to environmental change and will continue working no matter where you put the animals.

Rhesus macaques in research laboratories have adjusted to eating monkey chow every day of their lives, just as our domestic dogs have adjusted to eating, and seemingly liking, their dog food. Even though every food pellet rhesus macaques eat looks and tastes exactly the same as every other, each pellet gets closely inspected and looked at from all possible angles, and half of them are discarded for mysterious reasons no one has ever been able to figure out. Maintaining their own idiosyncratic feeding routine probably helps rhesus macaques keep their sanity and fight the boredom of life in a captive enclosure. Unlike many other wild animals, which become lethargic and depressed in the enclosures of zoos or research facilities, rhesus macaques are always alert, active, and seemingly in good spirits.

There is one thing, though, that rhesus macaques can't do without, and that is the company of other rhesus macaques. They are very gregarious animals and don't do well at all on their own. As many experiments done in the 1960s and 1970s proved beyond any possible doubt, when rhesus macaques live in isolation from their group members, they just go crazy.[14] Rhesus macaques would give up their food and water to be near other rhesus macaques—especially their family members—and be able to huddle together and groom each other. They just love the feel of another warm body next to them and the chance to comb and pick through its fur with their fingers. A kitten living around the SubDepartment of Animal Behaviour of the University of Cambridge, where I once worked, had figured out that he could get hours of free body massage and grooming from the rhesus macaques by lying against the fence of their cage. The macaques would fight one another for the chance to groom him.

As long as they are allowed to live in large groups, rhesus macaques organize themselves and behave as if they were in their native Asian forests. They become oblivious to what goes on around them and concentrate instead on playing their Machiavellian social games with one another. I have studied rhesus macaques in Europe, North America, and the Caribbean. They are Machiavellian everywhere they are.

Chapter 3 NEPOTISM AND POLITICS

Universal Nepotism

Politicians, religious and military leaders, wealthy people, and anybody else with any influence in public life use their power to help their family members advance in their careers, make money, or generally have a comfortable and successful life. The more there is at stake, the more people try to help their kin. In Italy, where I grew up, nepotism didn't surface in an individual's public life until early adulthood. Powerful parents didn't bother to get involved in their children's grades or class rank in school. Later, however, when competition for money and jobs began, nepotism became the overriding factor. Family pedigree was by far the best predictor of professional success. The lives of young people who up until then had run on parallel tracks were suddenly jolted upward or downward depending on the power, or lack thereof, of their parents.

In Italy, if college students decide to pursue a career in academia, their chances of success depend almost entirely on whether or not their parents have enough influence on the academic system to open its doors for them. Undergraduate students whose parents are politicians or well-established academics themselves have the opportunity to do their research training under the supervision of a powerful professor because their parents will pick up the phone, call the professor, and ask that their children be taken as advisees. As a Ph.D. student, I worked in a research laboratory where most of the other students, postdocs, and senior researchers were the sons and daughters of politicians and professors. My advisor turned down students without the required family pedigree even if they were academically outstanding and he had empty slots in his lab. He had to do it because his phone could ring at any time with a request to take a student he couldn't turn down. Every now and then, somebody without the right family pedigree makes it through the cracks in the system, but they typically don't get very far or last very long.

After undergraduate students get through the golden gate of academia with the help of their family members, they are essentially adopted by their advisors and become part of their extended family, thus entering another nepotistic system. They become Ph.D. students and later postdocs with the same advisor and remain in their mentor's shadow for years and years until their loyalty is finally rewarded with a permanent position. Students and postdocs spend as little time away from their advisors as possible because their future careers depend on the strength of their personal bonds with their advisors, and these bonds must be constantly attended to and nurtured. When powerful professors—called "baroni," to underscore this medieval organization of power—form committees to hire researchers or new professors, they negotiate with one another over which of their adoptive family members will get the job before the job is advertised and applications are reviewed. Getting to the very top of academia, however, or to the top of any other profession or business in Italy, requires entry into a third nepotistic system, that of political parties. When the stakes are that high, membership in one of these "families" is strictly required to sit at the table and play the game.

Italians are also well known for other forms of nepotism and nepotistic organizations—the Mafia, for example—but didn't really invent any of them. How overt or subtle nepotism is in public life may vary across human societies, but people are biased in favor of their kin in every human society. Nepotism is not just part of human nature; it's part of animal nature. Some animal societies are more or less nepotistic than others, but there is no society in which individuals are biased in favor of nonkin and against their kin. The reason nepotism exists is a phenomenon called kin selection. By helping relatives with whom they share genes, individuals increase the probability that their own genes will be passed on to the next generation. It is remarkable that the correct explanation for this basic aspect of all animal social life was first understood only in the mid-1960s, thanks to the intuition of a British biologist named William Hamilton.[1]

Without information on kinship, it is virtually impossible to understand how any animal or human society is organized and why the individuals in it behave the way they do. The basic structure of rhesus macaque society was first understood only in the mid-1960s, when researchers on Cayo Santiago started keeping track of maternal relatedness among individuals. In rhesus macaques, females mate promiscuously with several different males. Adult males typically don't know who their offspring are and don't provide any parental care. Paternal relatedness in a rhesus macaque group can be unequivocally assessed only with genetic testing, but this type of analysis has only recently become available in primate behavior studies. Maternal relatedness, however, can be reliably established from observations of births and subsequent association between mothers and infants.

As maternal genealogies on Cayo Santiago began to be assembled with long-term observations spanning several generations, rhesus macaque social behavior began to make sense. What looked like random acts of kindness between individuals turned out to be shameless acts of nepotism. We now know that by looking at who is sitting or walking next to whom, one can get a snapshot of the kinship relations within a group, because relatives always hang out together—at least female relatives do. Nepotism applies less to males than to females in rhesus society. To understand why and, more generally, to know how nepotism shapes the organization of any society and is shaped by it, one needs to know how relatives of both sexes are distributed in space and time. And how do we know that? Enter Sigmund Freud.

Incest and Dispersal

Freud was convinced that sex between family members explains a great deal about human behavior. He believed that people, and especially children, have a natural impulse to have sex with their family members: boys crave sex with their mothers (the Oedipus complex), girls crave sex with their fathers (the Electra complex), and both boys and girls crave sex with their opposite-sex siblings.[2]

To control these natural impulses—the story goes—people have created cultural taboos against incest. Despite the facts that children are taught that incest is bad and that adults are sanctioned by society if they engage in incest, the desire for incestuous relationships—according to Freud—is so strong that it pops up in every aspect of our social life. In his view, many psychopathologies can be related to these impulses.

Freud was a man with brilliant intuitions who, unfortunately, was way ahead of his time. His intuition about the importance of incest for behavior was right, but he got the facts all wrong. The way things really work is that many animals, including people, have a natural tendency to avoid sex with family members. Breeding with relatives, called inbreeding, is a bad thing for simple genetic reasons. People carry two copies of their genes, one inherited from their mother and one from their father. These genes exist in many variants, called alleles. For example, the gene for eye color exists in several alleles, which result in blue eyes, brown eyes, green eyes, and so forth. Some alleles have deleterious effects on survival and reproduction. However, most of these bad alleles are recessive, which means that when they are paired with a good allele, their effects are not expressed. The effects of the bad alleles are expressed only when they are paired with another bad allele. Recessive bad alleles are generally rare in the population. So, if we mate and have a child with an individual at random in the population, the probability that this individual has the same bad allele as we do—resulting in our child inheriting two copies of it—is very low. Because our family members carry many of the same alleles we do, however, having a child with a family member dramatically increases the probability that the child will inherit two copies of any bad recessive allele we might have. That is why there is a high incidence of infant mortality, sterility, and all kinds of genetic defects or physical malformations in populations in which individuals breed with their relatives.

People typically know who their relatives are because as children they are raised in small nuclear families and are introduced to other relatives or shown their pictures. Relatives may also be identified by the fact that they often share the same name. Some of these options are not available to animals, so many animals identify their kin using a rule of thumb based on familiarity: the individuals you grow up with are likely to be your kin. Therefore, they don't make good mates. A similar effect of familiarity on incest avoidance also exists in humans, where it's known as the Westermarck effect, after the sociologist who first noticed that people who had extensive contact with each other during childhood rarely married. The Westermarck effect has been demonstrated in many human societies around the world.[3]

Breeding with relatives is so bad that not being sexually attracted to individuals who are known, or likely, to be family members is not enough. Mistakes could still occur, and the price for them would be very high. So many animals are preprogrammed to take a further step to avoid inbreeding: they move away from their families before they begin reproducing. This phenomenon is called dispersal. Keeping track of who is leaving home and where they are going, however, could be very complicated. Therefore, in many animal species, dispersal is regulated by a simple rule: the individuals of one sex emigrate from their natal group, while the individuals of the opposite sex remain. The tendency to remain in the place where one is born is called philopatry, which in ancient Greek means "love of your homeland." A clear difference between males and females in dispersal and philopatry is very common in mammals. In rhesus macaques and most of the other primates, it is the young males that emigrate, while the females stay with their mothers. Rhesus males have a biological predisposition for emigration that appears early in life. In case they don't listen to their instincts, they get strong encouragement from their group members to emigrate when the time has come: they get kicked out of their natal group.[4]

There are multiple and complex reasons why in some species males disperse, in others females do so, and in a few species both males and females emigrate.[5] In rhesus macaques, males disperse because it is less costly for them to do so than it would be for females. Female dispersal would probably be a risky business. If females left their natal groups and spent lots of time alone, they would be very vulnerable to predators, they would be at risk of being attacked by other rhesus macaques, their infants might be killed, and it would be hard for them to find food and defend it from other individuals. When males disperse, they face some of these risks, but not as many as females would. In addition, staying with female relatives and sharing resources with them is not so bad for rhesus females, whereas it would be difficult to do for males. The balance between the costs and benefits of breeding within one's own group versus the costs and benefits of dispersal can vary from species to species, and that explains why there are species differences in dispersal.

Nepotism, Politics, and the Origins of Same-Sex Bonding

Inbreeding avoidance and sex-biased dispersal are key elements for understanding the social organization of a species. The result of male dispersal in rhesus macaques is that groups are composed of several clusters of female relatives with overlapping generations, called matrilines. In each matriline there are adult females with their mothers, grandmothers, sisters, cousins, daughters, granddaughters, and other female relatives. The only adult males found in a group are males that have immigrated from other groups, typically independently from one another. Therefore, the average adult female in a rhesus group is surrounded by relatives, while adult males have none. Because of kin selection, females form strong social bonds with their female relatives and help one another, whereas males can only count on themselves and any friends they can make. Matrilineal relatedness and strong bonds between related females are the glue that keeps rhesus macaque societies together. For this reason,

FIGURE 4 The glue that keeps rhesus macaque society together: Grooming between mother (*left*) and daughter (*right*). (Photo: Stephen Ross.)

rhesus macaques and other primate species with a similar social organization are called female-bonded species.[6]

When individuals behave in a way that benefits their kin, they increase the probability that copies of their genes will be passed on to the next generation through their kin's offspring. That's the reward for being altruistic to relatives, and most of the time, it's a good enough reward to justify good family values. When individuals help nonrelatives, however, they typically expect something in return. In Machiavellian species like rhesus macaques, there is no such thing as a free lunch. Social interactions between unrelated individuals are business transactions in which services are provided in exchange for other services. These business transactions are regulated by the laws of supply and demand. So, while altruism with relatives is called nepotism, altruism with nonrelatives is basically politics. The functioning of rhesus macaque society as a whole is therefore accounted for by a combination of nepotism and politics. Nepotism is a distinctive feature of female social relationships, whereas all the males can do is politics.

In a female-bonded and female-dominated species like the rhesus macaque, a few adult males are tolerated within a group because of the services they provide. Contrary to what one might think, sex is not one of them. Rhesus females are interested in sex in only six months of the year, and then only a few days each month. When they are interested, they can get all the sex they need from males who briefly visit them from other groups. Rhesus females do have sex with the males who reside in their own group, but only because this is the price they have to pay to keep those males around. So what do females get from the males? In one word: protection. Since rhesus males are bigger and stronger than females, they can protect them and their infants from predators as well as from other rhesus macaques.

This may come as a big blow to some male egos, but females don't establish and maintain a long-term relationship with a male so that they can have sex repeatedly with him. Instead, it's the other way around. Females consent to having sex repeatedly with the same male in order to establish and maintain a long-term relationship with him. Females can reproduce successfully through casual sex with a stranger—and in some cases, they are very happy to do so. Since females can have the sperm of almost any male they want any time they want, when they associate with a male, they are typically looking for something other than sex: investment in themselves and in their offspring. Rhesus males do not engage in parental care; therefore, the only investment they provide is protection.

Male dispersal and female philopatry are the predominant primate trend, being found among most species of monkeys. The great apes, however, have diverged from this trend. Chimpanzees, and to a lesser extent bonobos, have male-bonded societies in which males are philopatric and females emigrate. In gorillas and orangutans, both males and females disperse. According to some anthropologists, humans behave like chimpanzees and bonobos: males stay and females emigrate. There are exceptions, of course. I left my country and dispersed across the Atlantic Ocean, whereas my sister and her family settled down a few blocks away from my parents' home. Aside from a few rhesus macaques who live in our societies disguised as human beings, such as myself, the anthropologists are correct: most human societies around the world are characterized by female dispersal and male philopatry. So our hominid ancestors probably started out like chimpanzees: they lived in groups with strong bonds between related males and weak male-female and female-female bonds. At some point, however, things changed: there was a strengthening of bonds between males and females, leading to monogamous sexual relationships, as well as an elaboration of male-male bonds to include the formation of alliances with males of other groups.[7]

One unique characteristic of human societies is that both men and women tend to maintain ties with their kin after they relocate. Because of language, we are unique among animals in our ability to maintain social relationships with others in the absence of physical proximity. Maintenance of social bonds with relatives who moved to other groups probably favored the evolution of alliance formation between groups. In some contemporary human societies, alliances between males from different groups are also cemented by the systematic exchange of women as wives.[8] The elaboration of male alliances—especially alliances between groups—was probably a critical feature of human social evolution. The formation of these intergroup alliances now sets us apart from all other primates and, more generally, all other animals, in which this phenomenon is virtually nonexistent. Clearly, as with rhesus macaques, the functioning of human societies is still accounted for by a combination of nepotism and politics, but the way we play the game is slightly different from the way other primates do it.

Cheap Tolerance for Relatives

Rhesus macaques don't help their sons and daughters through school or college, but they try to make their lives easier in other ways. In rhesus society, every benefit comes with a price tag attached to it, but kin get discounts. A case in point involves social tolerance of other individuals. Rhesus females allow their family members to be near them—to walk, sit, eat, and sleep close to them—but don't make them pay a price for it. Social tolerance of relatives sounds like a trivial form of nepotism, but in rhesus society it's a big deal. Here's why.

Rhesus macaques are irritable creatures who have a low threshold for aggression. They wouldn't make it in the world without their fellow macaques, but they have very few inhibitions against attacking and hurting one another. In a society in which everybody is armed and dangerous, being close to another individual can be risky business. You must watch where you're going at all times as well as monitor everybody else's movements to make sure you don't find yourself in somebody's path. Sitting next to someone else to eat your lunch can be an offense grave enough to trigger aggression. Eugene Marais, a South African journalist who studied baboons a century ago, once wrote, "It was not long before we came to realize that the life of the baboon is in fact one continual nightmare of anxiety."[9] Well, rhesus macaques may have even more nightmares than baboons, and in most of these nightmares, Freddy Krueger probably looks like one of their fellow group members.

People have seemingly adapted to a life surrounded by unrelated and unfamiliar individuals. Yet the notion that being physically close to a stranger is dangerous—particularly in restricted spaces where opportunities for escape are limited—seems to be so engrained in our brains that we automatically take preemptive action. When riding in an elevator with strangers, people do their best to stand still, not touch others, and avoid mutual eye contact; that is, we avoid anything that may trigger aggression, even though the probability of being the victim of a deadly attack in an elevator is virtually zero. People will not occupy a seat next to a stranger in a movie theater if more distant seats are available. Men space themselves out when standing in front of a public urinal. We could interpret these behaviors as expressions of protection of personal space, but that would only be a fancy way of saying protection from the risk of aggression entailed by proximity. Our lives may not be a continual nightmare of social anxiety, but we've all had our Freddy Krueger dreams. The fear of being in proximity to others was probably in the brains of our primate ancestors long before they split from the ancestors of rhesus macaques. I doubt this fear will disappear from either species anytime soon.

In Machiavellian primate species, social life is inherently dangerous, but avoiding everybody all the time is not a good long-term solution to the problem. Whether rhesus macaque or human, you can avoid others for only so long before someone comes knocking on your door to vent all of their anger on you. A better solution to the problem is obtaining someone's protection. Sitting close to a powerful individual can protect one from aggression because others will think twice before firing their weapons for fear of hitting the wrong target. Even being close to a low-ranking individual can be beneficial because a potential aggressor could decide to switch targets at the last minute and hit your neighbor instead.

Being close to the most dominant individuals in a rhesus group—the alpha male and the alpha female—can be the most dangerous or the safest place to be, depending on whether the alphas' attitude is belligerent or peaceful. Clearly there is a price to pay for their

FIGURE 5 Receiving grooming is relaxing, even with an infant sitting on your belly. (Photo: Dario Maestripieri.)

tolerance, and in many cases that price is grooming. To a rhesus macaque, having someone's hands clean and comb one's fur is probably a pleasant experience comparable to that of getting your hair done and receiving a body massage at the same time. When rhesus macaques are being groomed by another individual, they sometimes relax so much that they fall asleep. Dominant individuals occasionally tolerate other individuals around, but they charge them hefty grooming fees. To be able to hide in the alpha male's shadow, others can't just sit there and smile at him—they have to groom him until their fingers get sore. Family members, however, get discounts. The alpha female's relatives can sit next to her and be safe free of charge. Female relatives groom one another a lot, but their grooming is more relaxed and less businesslike than the grooming between nonrelatives.

Given the high proximity fees charged to nonrelatives, it is not surprising that rhesus macaques end up spending much more time near their family members than near nonrelatives. Mothers and infants spend more time together than any other pairs of individuals, but other female relatives—sisters, cousins, aunts and nieces, grandmothers and granddaughters—also spend a great deal of time together. Adult males typically spend a lot of time on their own; they sit alone, eat alone, and travel alone. They sometimes hang out with young males, but it's the young males who follow them around and do all the work to maintain the relationship. In most cases, there is a clear reason why an adult male is spending time close to another adult. It's either about sex or there is an ongoing business transaction—grooming is being exchanged for tolerance or help. Even so-called friendships between particular adult males and females are probably business relationships of some kind, but primatologists have not yet figured out the nature of the business. Adults without relatives in their group or with little social appeal (for example, because they are old, low ranking, and no longer attractive as mates or protectors) are typically socially isolated.

Social tolerance among relatives by no means entails complete immunity from aggression. Just as people do, rhesus macaques fire their weapons at their family members, and especially at them. In fact, the highest rates of aggressive interactions within a rhesus group are often between mothers and daughters, or between sisters. The explanation for this paradox is that tolerance is a relative measure and takes into account the opportunity for aggression. Tolerance is the rate of aggressive interactions between two individuals divided by the amount of time they spend together. Kin fight with one another a lot, but spend huge amounts of time together, and most of that time is spent without fighting. Nonkin fight less often, but are at one another's throats almost every time they come close to one another—that is, every time they get a chance. In addition, aggression between kin is not as severe as aggression between nonkin. Finally, rhesus macaques are more likely to make up after they fight with kin than with nonkin. Rhesus macaques, however, reconcile after only a small percentage of their fights.[10] Most of the time, they just hold a grudge. So even tolerance and forgiveness for relatives have their limits.

Altruism?

Providing low-cost tolerance to family members is an important expression of nepotism, but not one that entails a great deal of effort or risk. Rhesus macaques can do better than that: they are capable of real acts of altruism. Altruistic behavior involves a significant cost, or risk, to the actor and a benefit to the recipient. Rhesus macaques don't donate money to charities, but they help others through grooming and agonistic support. Grooming benefits its recipients by cleaning their fur and relaxing them. The groomers spend time and energy as well as lower their vigilance, thus putting themselves at risk from predators and other macaques. Agonistic support involves intervening in a fight between two or more individuals and helping one against the other(s). Receiving help clearly increases the chances of winning the fight, but giving help entails the risk of injury and retaliation.

Rhesus macaques groom and support relatives much more than nonrelatives. Mothers groom their offspring for hours and hours—sometimes without getting grooming, or anything else, in return. And mothers don't hesitate to jump into a fight to help their offspring, regardless of who the offspring are fighting with and whether they were the victims of aggression or the ones who started the fight. When offspring are attacked by a higher-ranking individual, mothers intervene by trying to call the aggressor's attention to themselves, thereby giving their offspring the chance to flee. Needless to say, mothers end up paying a high price for their distraction tactics. Young offspring won't be able to reciprocate the agonistic aid received from their mothers until they become adults, so for many years mothers put themselves on the line for their sons and daughters without getting anything in return. That's definitely more altruistic than donating money to a charity in exchange for a tax deduction. Yet rhesus mothers don't behave exactly like Mother Teresa of Calcutta.[11] The altruism of rhesus mothers is based on nepotism, whereas the altruism of Mother Teresa wasn't, unless she had many more children in her long life than she led anyone to believe. In rhesus macaques, grooming and agonistic support also occur between nonrelatives, but much less frequently than between mothers and offspring, and as a calculated investment that is meant to produce an immediate return. When grooming or support is offered to a nonrelative, the "altruist" is counting every penny of his or her investment and hoping to make a profit from the transaction.

Support from relatives—particularly mothers—is a crucial feature of female social relationships and one of the pillars of rhesus macaque society. Family pedigree is by far the best predictor of a rhesus female's success in life because the more powerful her family members, the more fights she can win. Family pedigree matters for males only until they emigrate from the group at puberty. When they join a new group, there will be no family members around to help, and the males' success will depend entirely on their physical strength, personality, and political skills.

Aside from grooming and agonistic support, there is little else rhesus macaques do to help others, whether kin or nonkin. Rhesus mothers don't even share their food with their infants. A mother will smack her infant in the head if the infant shows too much interest in her food while she is eating. In fact, it is not unusual for a mother to take food out of her infant's mouth to eat it herself. The infant could be starving to death, but it wouldn't matter. All the altruism rhesus macaques are capable of has been preprogrammed into their brains,

FIGURE 6 Female relatives stick together: A family scene on Cayo Santiago. (Photo: Dario Maestripieri.)

and food sharing is simply something the software developer didn't think about when the rhesus computer was assembled and released on the market.[12]

Making Life Harder

To a rhesus macaque, the world of nonkin is divided into three categories: the individuals with whom one can have sex, the individuals with more power with whom one can do business, and the individuals with less power who can be victimized. Knowing rhesus macaques, one would be tempted to think that they harass unrelated individuals with less power because it's fun and because they can. Although there may be some truth to that, a more scientific explanation for this behavior is that by harassing others, or as evolutionary biologists say, "by inflicting fitness costs upon them," rhesus macaques indirectly help themselves and their family members. The bottom line of social life is that although individuals gain benefits from living in a group, they also compete with other group members for resources, including food, water, space, and mates. Therefore, in addition to providing social tolerance and helpful behavior, there is a third and more Machiavellian way in which rhesus macaques can make their relatives' lives easier, and that is by making the lives of their competitors harder.

Infants are probably at the top of a rhesus female's list of the individuals who can be victimized. Rhesus females have a great passion for infants; they are a big theme in their lives, along with kinship, dominance, and sex. At the beginning of the birth season, life in a rhesus group revolves around the newborn infants. Adult males stay as far away from infants as possible; males, with the exception of the alpha male, are completely ignored by the females and essentially hang out on their own for six months, resting, eating as much as they can, and making plans for the next mating season. Rhesus females, however, constantly check out the new infants—they watch them, touch them, and make strange sounds to them all the time. Yet they treat the infants of other females quite differently from their own. I usually can tell whether an adult female is holding her own infant or the infant of another female just from the way she is holding it.

With the exception of young and inexperienced females who don't know how to hold infants properly or mothers who have the habit of maltreating their offspring, macaque mothers handle their own infants with great competence and care, while they tend to be careless or outright rough with the infants of other females. They pull them, squeeze them, drag them, and in some cases, hit and bite them. One possible explanation for this behavior is that by harassing unrelated infants, rhesus females make the lives of their own infants' competitors harder. Consistent with this explanation, when the group is very crowded, harassment of infants becomes more frequent.[13] Females sometimes end up killing other females' infants, but it's rare and in most case it appears unintentional. Why rhesus macaques waste their time harassing their competitors when they could easily kill them is an interesting question that will be addressed in chapter 5 when their wars and revolutions are described.

Unrelated infants are by no means the only category of powerless individuals that can be victimized. Rhesus females are happy to victimize any unrelated lower-ranking individual, regardless of age, whenever they get the chance. Their behavior might benefit their relatives, but they would behave this way even if they had no relatives in the group. The reason is that their closest relative—self—is always there and is always the one they're happy to help the most. Whether the genes rhesus females are trying to perpetuate happen to be in their own bodies or in those of their relatives and their offspring doesn't make much difference. The difference between selfishness and nepotistic altruism is one of degree and not of substance. Rhesus macaques are preprogrammed for both selfishness and nepotism, and the neurons that control these behaviors are probably not very distant from one another in their brains. These neurons have been there for a long time and are quite good at what they do. Just as selfishness does not require a sophisticated concept of self, nepotistic behavior does not require an understanding of the concept of kinship.[14] Concepts of self and kinship have appeared very recently on this planet; the animal kingdom was doing just fine without them, and rhesus macaques will continue to do fine without them, for a long time.

Young Nepotists

Rather than being told who their relatives are and why it's important to be nice to them, rhesus infants are simply immersed—virtually from the day of birth—in situations that almost inevitably favor the emergence of nepotistic behavior. Infants are always with their mothers, and because their mothers spend a lot of time near their family members, so do the infants. So infants start hanging out with their siblings, cousins, aunts, and grandmothers simply because these individuals happen to be around. In addition, they observe how their mothers behave toward these individuals and toward others, and they begin to learn who is good and who is bad, who is safe and who is dangerous. Mothers don't lecture their offspring on family values, but keep them physically close to themselves and their relatives and chase nonrelatives away from them. The result of all this is that infants develop social networks that mirror those of their mothers.[15] Their mothers' friends are their friends too. As relatives become more and more familiar to infants, infants develop social preferences for these individuals and begin to act on these preferences. At some point, infants cease to be passive inhabitants of their social world and begin to actively seek out the company of the individuals they know and like best. Social preferences for female kin become especially strong in females, whereas males develop preferences to be with other males. First they spend a lot of time with male playmates of the same age. Then they become fascinated by older males and begin to form relationships with them that will eventually lead to the breaking of their bonds with their families and to emigration from the group.

The development of kin preferences and nepotistic behavior could be the product of a simple rule of thumb: form and strengthen your bonds with the individuals you know best. In theory, however, monkeys could recognize their kin as the individuals that look most like them. This mechanism, known as phenotype matching, would imply that monkeys have some idea of what they look like, perhaps as a result of seeing their faces reflected in water or some other reflecting surface. Another possibility is that they recognize their kin from the way they smell. This would imply that they have some idea of what they themselves smell like, a phenomenon called the armpit

FIGURE 7 Male bonding: An adult male grooming a juvenile. (Photo: Dario Maestripieri.)

effect, after a metaphor in which the monkeys would sniff their own armpit to know what they smell like and then compare it with the smell of other individuals.[16] Recognition of kin through visual or olfactory phenotype matching, in addition to age proximity, could explain why, according to some recent studies, females hang out more with their paternal sisters—other females with the same father but a different mother—than with completely unrelated females.[17] Since the mothers of the young females do not associate with the mothers of their paternal sisters, bonds between paternal sisters don't seem to be the result of familiarity and previous association. For some reason, however, I have my doubts that rhesus macaques know what they look like. If they treated their relatives the way they treat their own images in the mirror, there wouldn't be much nepotism or female bondedness in rhesus society.

Chapter 4
AGGRESSION AND DOMINANCE

Between Cows and Spotted Hyenas

Our capacity to do despicable things unto others is well known and deeply rooted. In the Middle Ages, torture and murder were conducted on a mass scale and on a daily basis. In our modern and civilized societies, whenever law enforcement is temporarily suspended or hindered during times of conflict or natural disaster, looting, rape, and murder are rampant. Konrad Lorenz, the Austrian ethologist who won the Nobel Prize in 1973 for his studies of animal behavior, argued that aggression is a necessary and an inevitable aspect of animal social behavior.[1] Impulses to act aggressively build up in our bodies, like fluid filling a tank, and need to be released every now and then—channeled into nonviolent competitive activities such as sports. Play tennis every morning and you'll be a nicer human being. Unfortunately, things are not that simple. Not many behavioral biologists these days share Lorenz's views on aggression. Aggression is neither necessary nor inevitable, and there is no fluid filling up our tank.

There are many animal societies in which aggression and violence are rare or unknown. Cows are pretty peaceful animals, as are sea turtles and the mountain gorillas we see in TV documentaries titled "Gentle Giants." At the other end of the spectrum, there are animals like the spotted hyena, whose cubs to try to kill their littermates from the moment they are born.[2] The newborn hyena cubs haven't even seen the world yet because their eyes are still closed, but they are already biting their siblings on the face. Hyena cubs are born with sharp canine teeth and are preprogrammed for murder. People and rhesus macaques are somewhere in between cows and spotted hyenas, but a lot closer to the hyenas than to the cows. Why?

It all comes down to economics and the notion that aggression, much like anything else, has benefits and costs. Mother Nature knows how to spend her resources, and you find aggression in animals only when the benefits are greater than the costs. In some animal species, aggression is rare because individuals wouldn't gain much from being aggressive, and therefore it's simply not worth it. In other species, it's rare because although individuals would gain something from it, they would have to pay a high price. It's when aggression is very beneficial and the price is affordable that you see lots of it. It's Economics 101. But what are these benefits and costs of aggression, exactly?

Aggression has a lot to do with competition; that is, the need or desire of two or more individuals for the same thing. The benefits of aggression involve obtaining what we want. Many animals compete over food, mates, or use of space. Let's take food as an example. Fighting over food may be worth the price in some circumstances but not in others, depending on the meal. A cat can eat any mouse he catches if he fights and defeats other cats who try to take that mouse away from him. Rhesus macaques can eat all the fruit they find if they fight and chase other macaques away from the tree where the fruit is. In these cases, the benefits of fighting are potentially high because the winner takes all. But imagine a herd of cows grazing on a large pasture. A greedy cow who wants to eat a lot of grass might start fighting with every other cow in the herd, but what would be the point? There is grass everywhere and way too many other cows to fight. In this situation, the greedy cow would be better off simply trying to eat as much as she can without worrying about what the other cows are doing. And that is exactly what cows do, most of the time.

The main cost of aggression is the risk of injury to yourself or a family member during a fight. Is it worth fighting over food if there is a risk that you or a family member could be seriously injured or killed? While the benefits of aggression depend on what individuals fight about, the costs of aggression depend on how individuals fight. In animals with deadly weapons such as large and sharp teeth or claws, the costs of aggression may be higher than in animals without such weapons. Animals with large weapons need them for hunting their prey but are typically very careful about how they use them on members of their own species. Lions are formidable predators, but you don't hear much about adult lions killing other adult lions. There are plenty of exceptions, of course. For spotted hyena cubs, the benefits of eliminating a brother or sister when there is not enough of mother's milk for two are high, and as a result, cubs are born fully armed to kill.

So why are rhesus macaques and humans so aggressive? Neither rhesus macaques nor humans eat grass or leaves as their main diet. Rhesus macaques and people tend to compete for things that are worth fighting for, where the winner keeps it all. In addition, neither rhesus macaques nor humans are naturally equipped with dangerous weapons such as horns, claws, or sharp teeth. A rhesus macaque can hit or slap another macaque, and a person can insult or punch another person, at relatively low cost to themselves. When our hominid ancestors discovered that they could throw objects to hurt or kill others from a distance (that is, when they invented projectile weapons), they found a way of reducing the costs of aggression to themselves even further.[3] Fighting from a distance is far less costly than direct physical combat and allows individuals to reap the benefits of aggression while minimizing its costs. People with firearms are more likely to kill other people because shooting someone carries far less risk of personal injury than attacking someone with bare hands or a knife. Opponents of gun control in the United States argue that guns don't kill people; rather, people kill people. The simple fact that the cost of killing people with a gun is lower than the cost of killing them without a gun, "benefits" being equal, explains why people with guns are more likely to kill than people without guns. So it's the guns that make the difference. In our evolutionary history, learning how to kill from a distance greatly improved our success as predators. Hunting from a distance allowed us to catch big animals such as the mammoth, but has had some unfortunate consequences for the way we compete with members of our own species. History shows that humans probably have killed more members of their own species than any other animal species on this planet.

So people and rhesus macaques are aggressive because, for us and them, aggression is a valuable tool that's available at a low price. It's a hot commodity. Somebody must have figured that out already because, these days, human societies attempt to control the aggressive tendencies of their citizens by reducing the benefits and increasing the costs of violent aggression. Reducing the benefits is accomplished by not rewarding violence and by teaching others, through religious or moral indoctrination, that nothing good comes out of violence. The costs of aggression and violence are increased through punishment. However, when these mechanisms for the control of aggression break down, such as during wartime or a natural disaster, aggression and violence typically flare up. When parents or teachers are not around, even children kill one another, as was aptly described by novelist William Golding in *Lord of the Flies*.[4] All of a sudden, aggression is on sale and everybody wants to buy!

Not every individual reacts the same way in these situations, of course. Even within aggressive species such as humans and rhesus macaques, there is a lot of variation. Some individuals are more aggressive than others, some groups of individuals are more aggressive than others, and both individuals and groups can be more aggressive at certain times than at other times. Anthropologists like to say that some human societies are very aggressive while others are very peaceful, so that it's difficult to generalize about whether humans are aggressive or nonaggressive as a species. Apples come in all kinds of varieties too; they differ in size, color, texture, and taste. Yet, as a whole, apples are different from oranges. Aggressive and nonaggressive species are as different from one another as apples and oranges. Aggression and violence have been reported in all human societies at various points in history. The same goes for every group of rhesus macaques that has so far been studied, whether in India or in the Caribbean, in Europe or in the United States. The existence of a truly peaceful human or rhesus society is as real as the existence of an apple that looks and tastes like an orange.

While the expression of aggression can be controlled, aggressive tendencies are an integral part of human nature. Aggressive tendencies are also an important facet of rhesus macaque nature, and because rhesus societies lack psychological indoctrination or institutionalized sanctions against violence, aggression is always latent in macaque social life and can be triggered very easily. So, when it comes to people and rhesus macaques, perhaps Konrad Lorenz wasn't so wrong after all. However, I wouldn't say that human aggression is necessary or inevitable, only that it's an "economically" profitable way to obtain what we want. To control it and reduce it, we have to make it less so.

What's All the Fighting About?

If aggression is about competition, there should be more fighting over scarcer resources, be they food, mates, or land. When there is plenty for everybody, there is no value in aggression, and none would be expected. Yet it doesn't seem to work that way. People and rhesus macaques fight even when there seems to be little to fight about. When things go well, both people and rhesus macaques appear incapable of sitting back, relaxing, and enjoying the peace. Instead, they concentrate their efforts on making one another's lives miserable.

When rhesus macaques are fed by people, so that they all get the same food and their bellies are always full, they fight even more than they would if they were starving in the forest. My research assistant Nancy was paid a decent salary to sit and watch rhesus macaques all day (actually, she did a lot more than that). As much as Nancy loved the monkeys and enjoyed working with them, she ended up quitting her job because she could no longer stand watching rhesus macaques pick on one another all the time. Their continual fighting seems to have no cause or immediate consequence. Aggression doesn't seem to come from competition, provocation, or miscommunication. The attackers don't seem to gain anything from their aggression other than stressing out their victims. This type of aggression is not some kind of pathological behavior or the result of an artificial environment. When rhesus macaques live in enclosures and are fed by people, they are probably bored and have more time available for social activities than they would in the forest, where they must travel and look for food. Being bored and having more time on your hands, however, are not good explanations for unprovoked aggression.

The seemingly senseless aggression of rhesus macaques makes perfect sense, however, if one realizes that what the monkeys really want and fight for is not this or that resource, but Power with a capital P. Power allows them to get anything they want. The struggle for power doesn't change with the amount of food on your plate or in your belly. Instead, it goes on day after day, week after week. If the monkeys are not busy traveling around and looking for food, that means they have more time to fight for power, and so they fight more. They are making good use of their time.

In theory, individuals without power could fight to gain power, and individuals who already have power could fight to maintain it or increase it even further. In rhesus macaques, however, the structure of power is very stable, and aggression typically runs in one direction, from top to bottom. With some notable exceptions, fighting is used on a day-to-day basis to maintain power and the status quo much more than to subvert it. As people with power know well, frequent and unpredictable aggression can be a very effective form of intimidation. Oppressive regimes and dictators around the globe use these tactics to maintain their grip on the societies they control. Even in democracies, induction of fear among people without power is used by the dominant classes and politicians to maintain their control over society. Rhesus macaques are very good at this. But what exactly is power in a rhesus society, and how does it work?

Dominance

Monkey power is called dominance, and to understand dominance we must first understand that rhesus macaques, just like people, have social relationships. Robert Hinde, a prominent animal behaviorist from Cambridge, once made the important distinction between interactions and relationships.[5] An interaction is an event between two individuals in which one does something to the other. For example, Bonnie giving Clyde a kiss is an interaction. An interaction is a unique event that happens at a specific time. When Bonnie and Clyde kiss many times and rob banks together, they have a relationship, because anything they do together is affected by what they've done together in the past and can affect anything they do in the future as well. The key to relationships is that individuals remember what's happened between them in the past and have expectations about the future.

Rhesus macaques have a very good memory for what Bonnie did to Clyde and when and why. Rhesus macaques probably don't engage in constant ruminations about the future as we do, but they certainly have expectations about what others are going to do in the future based on what they've seen them do in the past. Each rhesus macaque has relationships, good or bad, with virtually all members of the group and has some knowledge of the quality of the relationships between other individuals as well. Typically, close female relatives have good relationships, maintained by hours of sitting together and grooming each other. On the other hand, females from different families have poor relationships, marked by a lot of avoidance or aggression and submission. Adult males rarely have good relationships with other adult males, but they sometimes make friends with particular females or with young males. Relationships can also be stable or unstable. Relationships between related adult females are typically stable, whereas relationships between adult males, or between immatures and adults, can be unstable.

Each rhesus macaque in a group knows which relationships are good and bad and which are stable and unstable. The social life of a rhesus group is like an endless soap opera. Everyone in the group tunes in day after day, sitting on the monkey couch and eating popcorn, keeping track of the plot as it slowly unfolds with all its twists and turns, and trying to guess what the characters are going to do next. Much like any soap opera aficionado, each rhesus macaque knows which monkeys are family members, friends, or enemies, who has a crush on whom, and who's sleeping with whom. Macaques have their own taste in terms of soap operas, though, and seem to prefer stories of fighting and fear over those of love and betrayal.

Imagine that rhesus males Clint Eastwood and Chuck Norris wake up one morning and decide they want to eat the same apple for breakfast. Given the names they carry, they both act tough and end up having a fight. The Texas Ranger knows his martial arts, but Dirty Harry has his .44 Magnum, so Chuck goes ahead and makes his day. Clint wins the fight and eats the apple. Suppose the next day Clint and Chuck want the same apple again. They both remember the fight they had the day before, so Clint is confident he can win again, while Chuck hasn't forgotten the Magnum and is afraid he'll lose again. Clint gives Chuck the look that made him famous in the classic Sergio Leone films, and Chuck shows that he's afraid even though Clint hasn't touched him. After this exchange of signals, Clint takes the apple without even pulling his Magnum. As days go by, Clint doesn't even bother to give Chuck the look anymore; he just goes for the apple. Chuck continues to be afraid of Clint and simply gets out of the way when Clint comes close.

Clint and Chuck now have a relationship, just as Bonnie and Clyde do, although they don't kiss and rob banks together. They have a particular kind of relationship, known as a dominance relationship: Clint is dominant and Chuck is subordinate. Although they probably won't fight over the apple again, there will be other fights between them. Clint will attack Chuck every now and then, for no reason other than to maintain the status quo and "remind" Chuck that he has reason to be afraid of him. In response to Clint's behavior, Chuck will avoid being close to him and will show signs of fear every time they get close to each other. Suddenly the scene looks more and more like a Sergio Leone film, in which conflicts get settled by an individual's cool, intimidating presence, which elicits fear without much gunfighting. One day, however, Chuck will pull a gun on Clint, and on that day there will be trouble in the monkey OK Corral.

Anybody, whether monkey or human, who has been watching Clint and Chuck act in this movie can easily tell who's dominant and who's subordinate and can therefore predict their future behavior. Dominance and subordination can be assessed in many ways, by observing who gets the apple and who doesn't, who threatens and who's afraid, who attacks and who flees. Since Chuck will always keep an eye on Clint and monitor every move he makes, one can tell who is dominant and who is subordinate in their relationship simply by their looking behavior. Back in the 1960s, some researchers suggested that one could understand all dominance relationships within a monkey group by simply measuring looking behavior; that is, what they called "the structure of attention."[6] The idea is that the individual who receives all of the attention is the most dominant monkey in the group, whereas the individual who is ignored by everybody else is at the bottom of the dominance hierarchy, and all the others fall somewhere in between. The relation between social attention and dominance is interesting and applies to people as well. Children and adults have dominance relationships very similar to those of rhesus macaques, and anyone who observes a group of children interacting on a playground or a group of adults sitting around a dinner table can gain some insight into their dominance relationships simply from their looking and monitoring behavior. However, things are sometimes complicated by differences in temperament or personality that may or may not be associated with dominance. There may be dominant people who are anxious individuals and monitor others constantly, and subordinates who avoid looking at others altogether. These individual differences exist in rhesus macaques as well and make it difficult to assess dominance on the basis of looking behavior alone.

Technically speaking, dominance is a property of relationships and not of individuals. For example, Clint Eastwood could be dominant over Chuck Norris but be subordinate to all the other tough guys in Hollywood. In theory, dominance could also vary within a pair and be context-specific; for example, Clint could be dominant over Chuck when they both want the same food, whereas Chuck could be dominant over Clint when they both want to date the same attractive female. This context specificity doesn't appear in rhesus macaques, however. When Clint and Chuck have a stable dominance relationship, Clint can kick Chuck's butt any time, any place, and they both know it.

Each rhesus macaque has dominance relationships with all the members of its group. Individuals who are dominant over many others tend to act dominant in general; for example, they walk around holding their tails up, look very self-confident, and are generally assertive in their behavior even in novel circumstances. Clint Eastwood acts the same in all of his spaghetti Western or Inspector Callahan movies. So does Chuck Norris in every one of his films, to be fair to him. Individuals who are subordinate to many others tend to be fearful and act submissively in general. Fear of being attacked can become so engrained in the mind of a subordinate like Chuck that even if one day Clint is not around and the apple is sitting there waiting to be eaten, Chuck will not touch it. There have been studies in which the most dominant individuals were removed from a macaque group, but the subordinates still would not take the food that was being offered to them.[7] If you are a subordinate in a rhesus macaque group, you play it safe.

Even the safest of players, however, can grab a golden opportunity when it presents itself. If the dominant male takes a longer than average snooze, a subordinate male who is afraid to even approach a female, let alone make a pass at her, when the dominant male is watching, has a small window of opportunity. He can mate with many females and get them all pregnant at the speed of light. Similarly, if a dominant individual is sick or injured or shows any signs of weakness on any given day, a subordinate can take advantage of this and attempt to challenge him and reverse the dominance relationship.

Hierarchy and Ranks

In rhesus macaques, dominance is transitive from one relationship to another. This means that if Clint Eastwood is dominant over Chuck Norris and Chuck Norris is dominant over Steven Seagal, Clint is automatically dominant over Steven Seagal as well. Because of this, all individuals can be ranked in a linear dominance hierarchy, with the individual who is dominant over all the others at the top and the individual who is subordinate to all the others at the bottom. An individual's position in the hierarchy is called its dominance rank. Rhesus society is organized a lot like the U.S. Army, except that female soldiers are not sexually harassed by males and all homosexuals are out of the closet.

A group of rhesus macaques has separate dominance hierarchies for males and females, although the two hierarchies overlap and intersect in a complicated manner. The top-ranking male, or alpha male, is dominant over all males and females. He's the king and has absolute power over everybody and everything (although he might still have to work hard to mate with females). Unfortunately, as with all good things, alpha male status doesn't last very long. It's so good to be king that some other male will soon come along and knock the alpha male off his throne. Some of the other adult males are dominant over females, whereas others end up being subordinate to many of them. In theory, since males are larger and stronger than females, all males would be able to win fights with females if the fights happened in a dark alley with no one else watching. Fights between males and females, however, rarely happen without an audience, or without audience participation. Females are helped by their female family members, and thanks to this help, some of them are able to rank above the males.

The top-ranking female, or alpha female, is the queen of the group. She's dominant over all the females and also over the adult males, with the exception of the alpha male, by virtue of the support she receives from him. In other words, the king will intervene on behalf of the queen every time she gets in trouble with the other males. The queen will also help the king if he's challenged by resident males or by males who have just immigrated from other groups, unless she's gotten tired of him and wants to get rid of him. Therefore, the alpha male and the alpha female in a rhesus group are partners in power, like Bill and Hillary during the Clinton presidency, according to a book about the couple.[8] If the queen decided to switch alliances and support another male against the current king, that would mean the king's days are numbered. Although males do fight one another for dominance, the winner of these fights needs the stamp of approval of the females, and especially their queen, to be the top-ranking male in the group. Without the females' approval, a male's life can be hell. It's sometimes hell even with female approval.

Because females belonging to the same matriline support one another when they fight, they tend to have similar ranks. Therefore, there isn't just a hierarchy of females within a group, but a hierarchy of matrilines as well. All members of the top-ranking matriline, usually the largest in the group, rank higher than all the members of the second-ranking matriline, who in turn rank higher than all the members of the third-ranking matriline, and so on. On Cayo Santiago, the rhesus macaque groups consist of three to four huge matrilines, each with up to fifty individuals. In natural habitats such as the forests of India, there are probably more, but smaller, matrilines within a group.

Within each matriline, young females acquire a rank adjacent to that of their mothers, but remain subordinate to their mothers for life. Sisters are ranked in reverse order of their age, so that a female's youngest daughter ranks above all of her sisters, her second-youngest daughter is the second-highest-ranking sister, and so on. This pattern of dominance among sisters is determined by the fact that when sisters fight with each other, the mother always intervenes on behalf of the younger sister. Thus the youngest daughter is always supported against all of her other sisters. There are many possible explanations for the mother's intervention behavior, some simple and others very Machiavellian, as we'll see.

Altruism and Opportunism in Agonistic Intervention

If rhesus males Chuck Norris and Steven Seagal have a fight, chances are someone else will get involved. If Clint Eastwood intervenes to support Steven against Chuck, he is forming a coalition with Steven. Agonistic intervention and coalition formation are especially common between females, but males do it too. Individuals join fights on their own, but more often than not, they are encouraged to do so by one of the fighters.

Fighting monkeys ask for help from others all the time. If Clint intervenes to help Steven against Chuck, there is a good chance that it was Steven who asked for his help. In fact, chances are that Steven picked a fight with Chuck so that he could ask for help from Clint and hope to make friends with him.

When an aggressor asks for help from others, it quickly alternates between threatening its victim and turning its head and looking at others who may be willing to jump in. In the primatologist's jargon, this behavior is called "show-looking." While doing this, the monkey typically stands on four legs and holds its tail up. Holding the tail up works both to attract attention and as an invitation to other individuals to join the fight. There are also specific screams and squeals that aggressors use to attract others' attention and solicit intervention. Sometimes aggressors just look around for anyone who might be willing to jump into the fight, but often they solicit intervention from a specific individual. They do so by looking repeatedly at that individual and attempting to make eye contact with him or her, and in some cases by positioning themselves right in front of that individual with the back to him or her and the tail up. If the individuals being solicited are not interested or willing to intervene, they act as if nothing's happening: they avoid making eye contact, continue to mind their own business, or walk away from the scene. Rhesus macaques can be very good at feigning indifference.

When victims of aggression ask for help from others, they use the same behaviors as the aggressors, but with a lot more screaming. They probably scream from fear or pain, but with their screaming they accomplish two goals: they disorient the aggressor, thereby interrupting the aggression, and they attract the attention of other individuals and solicit their support. When victims of aggression scream, their screams sound different depending on whether they are trying to avoid being attacked again or trying to fight back (staring back or lunging at the aggressor) and on whether or not they are in a lot of pain from the attack. Researchers who believe that rhesus macaques have a mysterious coded language, not discernible to the human ear, think that the victims' screams also sound different depending on who the aggressor is and the circumstances of the aggression.[9] In this view, when victims of aggression scream, they are broadcasting a message over the airwaves that might mean something like "Alert! Alert! I am under attack by a subject with the following characteristics: sex: female; age: fifteen; category: nonrelative; family of origin: top-ranking matriline. Note: Subject is armed and dangerous. Note 2: I'm already in a lot of pain. Please help!" Whether or not the message reaches its intended recipient, whether or not it's properly decoded, and more importantly, whether or not the intended recipient gives a damn about it, depends on who the transmitter is, who they are fighting against, and who else is watching and listening. You might expect that everyone would ask for help from the king and the queen of their group, but that's not always a good idea, for a couple of reasons. First, like all royals, the rhesus king and queen are very good at feigning indifference when the requests for help come from somebody at the bottom of their society. Second, if they are really bothered by all the noise and insulted by being asked to get their hands dirty with fighting, they may intervene and attack the victim who's seeking help and not the aggressor.

Intervening in a fight on someone else's behalf is a form of altruistic behavior. Unless that someone is a family member, intervening involves a decision-making process with some selfish calculations. This process has been studied by many social psychologists, but with people, not monkeys. They call this type of work bystander intervention research.[10] Typically, social psychologists present a subject (the "bystander") with a hypothetical scenario in which he or she witnesses a dispute between two people, or a car accident, or any other situation that might require an altruistic intervention, and then the subject is asked under what circumstances he or she would be willing to help. The researchers then attempt to uncover the variables that affect the subject's decision making (for example, the identity of the individual in need, the risks and rewards of intervention, or the presence of other individuals who might contribute help or judge the subject's behavior). The social psychologists interested in bystander intervention research could learn something from the way rhesus macaques make decisions about joining a fight, but unfortunately most of them are not well acquainted with these monkeys.

Explaining agonistic intervention in rhesus macaques—or, more generally, any form of altruistic intervention in monkeys or people—may seem very complicated, but ultimately (if we leave issues of morality and religion aside), it all comes down to kinship and economics; that is, the costs and benefits of intervention. In rhesus macaques, kinship and nepotism explain agonistic intervention between family members. This is especially true for maternal agonistic support of offspring, which is one of the most common forms of agonistic intervention. Rhesus mothers consistently support their offspring, irrespective of their offspring's role as aggressor or victim or the rank of their opponents. When mothers help their offspring against higher-ranking opponents, they have no prospect of winning the fight but expose themselves to high risk of retaliation—and you can be sure that retaliation will follow. Yet mothers sacrifice themselves by screaming at or slapping the aggressors to direct attention to themselves and give their offspring the chance to flee. Mothers are the only real altruists in rhesus macaque society. Occasionally, other relatives will help one another too, but agonistic intervention between unrelated individuals is strictly politics.

The way adult males who have recently immigrated into the group intervene in fights epitomizes the opportunistic nature of agonistic intervention when kinship is not involved.[11] Since these males don't have any offspring in the group, they never intervene to help infants and juveniles—there is no point in doing so. Instead, they typically intervene on behalf of other adult males or adult females; that is, individuals who might be able to help them in return. Adult males also typically help aggressors, not victims. Therefore, they don't intervene to protect somebody who is in trouble. Instead, because aggressors are typically higher ranking than their victims, adult males usually help the individual who's going to win the fight anyway. They also tend to help individuals ranking higher than themselves who are fighting against lower-ranking opponents; that is, individuals who don't really need their help. Essentially, the adult males are not really helping; they are just sucking up to higher-ranking individuals at a minimal risk or cost to themselves. What they hope to accomplish is twofold: first, they hope to make friends with higher-ranking individuals and have their cheap help reciprocated; second, they hope to score dominance points with the victims of aggression. If the victims of aggression rank lower than the "helpers," the helpers will reinforce their dominance over them and maintain the status quo. If the victims rank higher than the helpers, by intervening against them, the "helpers" hope to outrank them in the hierarchy. This type of opportunistic intervention is commonly displayed by males who are actively jockeying for status, such as subadult males who have not yet emigrated from the group and new immigrant males. In both cases, these males are attempting to climb the male dominance hierarchy and exploit any available opportunity to outrank another male. When males intervene to help females, they may gain tolerance from both those females and their relatives. If they are lucky, they may even gain some sexual favors. Opportunistic agonistic intervention is also common when females help nonrelatives. The rules are: help when it doesn't cost you anything to do it, make friends with powerful individuals, and try to take advantage of any opportunity to consolidate or improve your position in the dominance hierarchy.

Scapegoating

When rhesus macaques are the victims of aggression, there is another way in which they attempt to get other individuals involved—this time not as helpers, but as scapegoats. They do this by immediately attacking another individual, or in the jargon of primatologists, "redirecting aggression" toward another individual. In a typical case of "redirection," a monkey who's just been attacked will immediately turn toward another monkey and chase, threaten, or directly attack that individual while looking back at the aggressor or other individuals and asking for their help.

The scapegoat can simply be a random individual who was minding his or her own business near where the first fight erupted. More often than not, however, the scapegoat is not chosen randomly. First, the scapegoat is typically someone who ranks lower than both the victim and the aggressor and who has no chance of getting help from anybody else, either because he or she has few or no relatives or friends in the group or because his or her family is so low ranking and hopeless that its members don't even bother to try to help one another anymore—in other words, a loser. Second, all the rhesus macaques in a group, except the monkey at the bottom of the hierarchy, have a favorite scapegoat, and whenever they are attacked, they will immediately look for their favorite scapegoat, even if he or she is not in the vicinity of the fight. Chances are, in fact, that the favorite scapegoat will *not* be in the vicinity of the fight because, knowing very well how the game works, he or she will take off and run as far away as possible every time his or her designated torturers get involved in a fight. Because the monkeys at the bottom of the hierarchy are the favorite scapegoats of many other individuals, they typically scatter all over the place whenever a fight breaks out within the group.

Scapegoating is really a Machiavellian strategy that accomplishes many things at once. At a basic level, finding a scapegoat is the victim's way of avoiding further aggression by diverting the aggressor's attention away from itself and toward another individual. It's the familiar "Why me? Hey, let's all go after this guy instead!" If they are successful in recruiting the help of their own aggressors or other individuals against the scapegoat, the original victims are also forming alliances with those individuals. Alliance formation through scapegoating can allow individuals to make friends with more powerful group members and reinforce their own dominance over the scapegoat. Very ambitious and Machiavellian macaques could even target a scapegoat who is higher ranking than themselves and use the situation to attempt to outrank that individual. This type of strategy is more likely to be used by adult males than by females because female dominance is constrained by the matrilineal structure of the group, whereas males' ranks are more flexible and may depend in large part on their social strategies, in addition to their fighting skills and their social and sexual charm.

Scapegoating can have yet another function by serving as a deterrent against future aggression from the same aggressor. If a fight breaks out between two unrelated individuals, the victim can redirect aggression against a relative of the aggressor, typically a juvenile or an infant.[12] In this case, redirection is really a form of retaliation against the aggressor's family member and functions as a warning that if future attacks occur, the aggressor's family members will suffer some painful consequences. This is where, in the rhesus world, Sergio Leone's spaghetti Westerns meet the Godfather, and where monkeys that give others the Clint Eastwood stare start talking in whispers like Marlon Brando.

A Social Climber's Story

Tequila was a two-year-old female belonging to the top-ranking matriline in her group. Tequila's mother, Yvette, was one of the alpha female's older sisters. Tequila spent most of her time hanging out with her mother and playing with other youngsters from her own and other families. Tequila liked to play rough, wrestling and chasing her playmates the way boys do. The older she got, the rougher she played. One day Tequila and her cousin Jemima wrestled harder than usual and ended up hurting each other. As usual, they screamed for their mothers. Both Yvette and Jemima's mother, Lola, rushed to the scene of the crime, each to help her own daughter. Yvette was Lola's younger sister and therefore ranked one notch above her in the dominance hierarchy. Yvette smacked Jemima and threatened Lola, who ran away screaming in fear. That put an end to the squabble between the girls.

That day, three monkeys learned a lesson, and one of them also had an interesting idea. Tequila learned that she could boss Jemima around because even if Jemima got upset and called her mother, Yvette would come and take care of them both. Jemima learned that if she got into a fight with Tequila she would end up losing because her mother was afraid of Yvette and couldn't possibly win a fight with her. Lola learned that from now on, any time her daughter played with Tequila there might be trouble, and that she would have to put up not only with her younger sister Yvette's bossy attitude, but with that of her bratty daughter as well. The monkey who had the idea was Tequila. Watching what happened between herself and Jemima, and later between her mother and her aunt, she realized that maybe from now on, she could boss around not only her cousin, but her aunt as well. A light went on in the mind of an ambitious social climber.

What happened that day between Tequila, Jemima, and their mothers later happened with Tequila's other playmates and their mothers. Tequila already knew who her playmates' mothers were, but now she also learned that if their mothers were afraid of her mother, she could be mean to the other girls and get away with it. She tested this idea by picking fights with other youngsters and then screaming for Yvette to come and help. It worked. Her playmates began to be afraid of her and to treat her with a great deal of respect. Good. One day, however, when Tequila picked a fight with one of the alpha female's young daughters, she learned that her mother was no match for the alpha female, and that it was her turn to be afraid of a playmate. Eventually, all the youngsters in the group, males and females, learned that the way their mothers treated one another would become the way they would treat one another too. Eventually, their mothers seldom bothered to get involved in their kids' squabbles any more. The kids started acting as if their mothers' shadows were always behind them, so they settled their disputes on their own depending on how large or small those shadows were.

The settlement of dominance relationships between playmates is only the first step in dominance rank acquisition. The integration of young macaques into the adult dominance hierarchy is not an automatic process, but requires work. The problem is that adults are larger and stronger than juveniles and very unwilling to let anybody enter the dominance hierarchy above themselves, regardless of who their mothers are. Rhesus infants enjoy a period of temporary immunity from adultlike aggression in the first three or four months of life, when all the other monkeys think they are cute and are willing to put up with them (but may also harass them and kidnap them, as we shall see). When this immunity ends, infants start getting threatened, slapped, and bitten just like everybody else. All of a sudden, infants become particularly vulnerable to aggression because they are small, weak, and inexperienced, and because you can be sure there is somebody in the group who has some beef with their family members. The end of the immunity period marks a painful transition in the life of a rhesus macaque. Youngsters start out being subordinate to all adults in their group and have to work their way up, if the status of their family members will allow it. The offspring of the lowest-ranking female in the group have no option but to learn that they can be victimized by everybody else and that things are unlikely to change. The sooner they learn about their future as everybody's favorite scapegoats, the better. The offspring of other females can aspire to achieve the rank of their mothers and gain the subordination and respect of all individuals that rank below their mothers. But they have to work at it, as nothing comes easy or cheap in rhesus macaque society.

And so it began. The day Tequila had the idea that she could boss her aunt Lola around, she started paying close attention to what happened every time her mother interacted with another adult. She started a file on every adult in the group and recorded in her mind who was afraid of her mother and who wasn't, who her mother treated with respect and who she didn't care for, who won the fights and who lost them. Soon the files were neatly stacked into two piles, one for adults who were afraid of Yvette and the other for those who were not. In other words, there was one pile for the monkeys above Yvette and one pile for the monkeys below her in the dominance hierarchy. When the files were complete, it was time for action. There wasn't much she could do about the adults in the higher-ranking pile, or so she thought at the beginning. But she could work on those in the other pile and try to get the respect from them that was due to her. She would target one individual at a time, and Lola, Jemima's mother, would be her first victim.

From then on, every day and many times a day, Tequila tried to pick a fight with Lola. She would get in Lola's personal space, threaten her, and then scream her head off to get her mother's attention and help. At first, Lola was just annoyed by Tequila's behavior. Obviously she wasn't afraid of Tequila and wasn't going to show this ill-mannered juvenile any sign of respect. She knew, though, that if she gave Tequila the beating she deserved, there could be trouble, so she just tried to act indifferent. When Tequila was very pushy, Lola would threaten her and even slap her. Tequila, however, was very persistent and became better and better at screaming at Lola and getting others interested in her show. Her mother came to her aid and threatened Lola, and so did other females from her family. The time finally came when Lola was worn out by this constant harassment and gave in. She couldn't really afford to fight Yvette and the rest of her family, so she decided to move aside and make room for Tequila above her in the dominance hierarchy. All Lola had to do was to show Tequila a sign of fear,[13] and one day, when she just couldn't take it anymore, she did just that. From that day on, Lola would be afraid of Tequila and treat her with the same respect she gave her mother Yvette and the rest of her family. The dominance relationship between Tequila and Lola was now settled once and for all.

Tequila moved on to the next victim, repeating the entire process with all of the adults who ranked below her mother. Eventually she was incorporated into the adult dominance hierarchy with a ranking just below Yvette's. Things were going so well for Tequila that when she reached the bottom of her first pile of cards, she thought she would pick a card from the other pile and see what happened. So one day she tried targeting a female who ranked above Yvette. All hell broke loose, and Tequila paid a high price for her impudence. Her mother knew Tequila was out of her mind to attempt this crazy stunt but tried to help her anyway, and both took a giant beating. Tequila put the other pile of cards in a drawer and hasn't touched them since. But she thinks about those cards every day, especially the one with the face of the queen on it. Maybe one day, when the time is right . . .

The process of rank acquisition by macaque youngsters can begin in the first or second year of life and continues into adolescence and young adulthood. It's basically the same for males and females. At puberty, the males leave the group. If they delay their departure, they become large enough to outrank their own mothers and some of the higher-ranking females, while remaining subordinate to older and larger males as well as to the alpha female. The sons of higher-ranking females sometimes assume high ranks within their own group, especially if they are big, and even manage to outrank some of the resident adult males. Achieving high rank in the natal group may delay a male's emigration. The sons of low-ranking females rank very low in the hierarchy, and the increasing amounts of aggression they receive are a good incentive for emigration.

Because juveniles end up ranking right below their mothers, they essentially "inherit" their mother's rank.[14] It's a clear example of social inheritance, however, and not genetic inheritance, as offspring don't inherit genes for dominance or subordination from their mothers. This conclusion is supported by the fact that juveniles that are adopted by unrelated females at birth acquire the rank of their foster mothers and not that of their biological mothers. In addition, when mothers rise or drop in rank, the ranks of their offspring also rise or drop.

The settlement of dominance relationships between sisters is an interesting part of the process of rank acquisition. Tequila, at some point, started picking fights with her older sisters and soliciting the support of her mother against them. Her sisters initially resisted these attempts at domination, but Yvette always sided with Tequila, and they were eventually outranked. Why mothers always help their youngest daughter against her older sisters has puzzled researchers for a long time.[15] Actually, one of my undergraduate students once had a very simple explanation for it: "It's because the youngest daughter is always the one who the needs the help the most!"

Another possible explanation for mothers' behavior is that since younger females can survive longer and produce more offspring in the future than older females, they have "higher reproductive value." Therefore, by supporting younger daughters against older daughters, mothers do their best to increase the probability that their own genes will be passed on to the next generation through the reproduction of their offspring. Another very Machiavellian explanation has to do with the mother's control of power over her offspring. Here it's important to know that just as mothers help their daughters against their sisters, the daughters, in turn, help their mothers when the mothers fight with their other daughters. So, by always supporting the youngest daughter against her sisters, mothers prevent their older daughters from forming coalitions with one another and attempting to outrank their mother. If mothers helped their older daughters, the power of daughters in the family would grow as the number of daughters increased, and they could eventually revolt against their mother. Instead, by always supporting the youngest daughter, the mother ensures that the power of the older daughters is reduced rather than increased with the birth of each female offspring, and that the youngest daughter will always help her mother if her older sisters attempt to revolt against her because it's in her own interest to do so.

The matrilineal structure of rhesus macaque groups and the way in which mothers always help their offspring against others ensures that the process of dominance acquisition follows a predictable pattern and that the ranks acquired by youngsters are themselves very predictable and stable over time. The manner in which unrelated individuals join fights also contributes to the establishment and maintenance of the matrilineal rank system. Unrelated individuals intervene only in fights they can win—that is, on behalf of the dominant individual of the pair—and when youngsters begin targeting adults that rank lower than their mothers, they and their mothers receive the support of unrelated individuals as well. Conversely, if youngsters were too ambitious and challenged adults that ranked higher than their mothers, as Tequila tried to do, other unrelated individuals would intervene against them. Thus intervention from unrelated individuals is important for the maintenance of the status quo. The youngsters are active players in this game as well, because they also opportunistically join fights initiated by unrelated adults against individuals that rank lower than their mothers or use them as scapegoats as a way to outrank them. In other words, they use all the Machiavellian tricks at their disposal to rise in rank and eventually attain the position in the dominance hierarchy where their family pedigree deems they belong.

Female dominance hierarchies are very stable over long periods, so females are likely to die with the same rank they were born with. Cases of rank change involving adult females who are already well integrated in the dominance hierarchy are rare, but they do occur. Such cases often involve rank reversal between mothers and daughters, or between sisters. Although quarrels between mothers and daughters end, as a rule, with the mother slapping the daughter and the daughter screaming in fear, the rule can be broken. Such a rare event may happen when a mother is old or sick and too weak or tired to put up with a rebellious daughter. The rebellious daughter may also enlist the help of the alpha male or other individuals against her mother and win a fight with her. Dominance rank between sisters can sometimes be reversed as well, as, for example, when their mother dies. Without the mother's presence and her constant support of her youngest daughters against their older sisters, the older sisters can challenge and dominate their younger sisters by virtue of their larger body size and strength, greater experience, or because they have more daughters of their own to help with fighting. In most cases, however, the death of an old matriarch does not automatically result in a dramatic change in family dynamics. Daughters and other family members simply remain in their place and are respectful of the order Mom established when she was alive. There is a lot of inertia in the matrilineal dominance system, and the ambitious and reckless individuals who decide to fight the system and change the rules have the odds stacked against them.

Life at the Top, and at the Bottom

Being high ranking or low ranking in a rhesus macaque group is not unlike being rich or poor, or powerful or powerless, in any human society. When survival is at stake, being rich or poor can make the difference between life and death. Clearly not all citizens of New Orleans were equally affected by Hurricane Katrina. The elderly and the poor perished in greater numbers. In rhesus society, dominants always travel in business class and subordinates in economy, and if the flight is overbooked, it's the subordinates who get bumped off the plane. The subordinates are always the last to eat and the first to be eaten. If there isn't enough food for everybody, the dominants feast and leave nothing for the subordinates to eat. If a tiger attacks, chances are low-ranking individuals are going to get eaten first because they are often relegated to the periphery of their group or sleep in places that are not safe.

When conditions are not so extreme that rank makes the difference between life and death, it can still make the difference between a good and a bad life. Because low-ranking monkeys are the target of more aggression, they risk being injured more often. Even if they manage to avoid being attacked, they must constantly watch their backs and monitor others' every move. Low rank can be synonymous with chronic stress, but there are exceptions. Some high-ranking macaques stress themselves out trying to maintain their power and control others with an iron fist. Similarly, some middle-ranking animals are very stressed by the aggression they receive from above, by the need to control those below, and by their own ambitions and efforts to climb the hierarchy. Conversely, there are low-ranking animals who seem to accept their condition without showing dramatic signs of stress. In general, unpredictability is stressful, whereas predictability offers opportunities for coping. Therefore, individuals with unstable rank are more stressed than individuals with stable rank positions, regardless of where they are in the hierarchy. Primate biologist Robert Sapolsky has shown this to be true for baboons as well.[16]

High-ranking macaques, nevertheless, can compensate for the stress of being dominant with the privileges and the commodities that are available to any member of the upper class, whereas no such option is available to the subordinates. The reward of power for those who command it is the opportunity to enjoy what they want, when and how they want it. So what is it that rhesus macaques want? Good food to fill their bellies, sex with the individuals of their choice at the time of their choice, hanging out in the cool spots (for example, sitting in the shade on a hot summer day, or sleeping in the most comfortable site available), many healthy and bratty infants to raise, and the constant attention and grooming efforts of other individuals. If a new show is playing, the high-ranking animals want first-row seats. If a new toy is discovered, they want to play first. They claim possession of anything the low-ranking monkeys may possess, and they want to be able to take it away from them at their pleasure. Finally, and most importantly, they want the prerogative to intimidate and harass their subordinates whenever the mood strikes.

For the most part, high-ranking rhesus macaques do obtain what they want and successfully claim and monopolize objects, locations, or other individuals at their whim. While some of these may look like trivial and ephemeral pleasures, they have substantive consequences. For example, because the young females born in high-ranking matrilines generally eat more and better and are less stressed, they reach puberty earlier than the daughters of low-ranking mothers. High-ranking females often produce a new infant every year, while low-ranking females produce an infant every other year. High-ranking males can have sex hundreds of times during the course of the mating season, whereas some subordinate males may not get any at all. Finally, high-ranking individuals may live longer than subordinates.

Low-ranking macaques are often condemned to eat leftovers, but more often than not they manage to obtain their slice of the pie by using so-called "alternative strategies." That is, instead of competing directly with the high-ranking animals over what they want, they come up with creative and circuitous routes to get what they need. For example, low-ranking monkeys learn how to be quiet and not tell anyone when they find food, or to eat quickly when the dominants are not looking. Subordinate males learn to hide behind bushes and out of sight of the alpha male to have sex, and subordinate females kidnap, harass, and torture the infants of dominant females when their mothers are busy grooming the alpha male. As a consequence of these alternative strategies, some low-ranking individuals manage to have healthy lives and produce as many offspring as the dominants, if not more. Moreover, subordinates are always ready to take advantage of any opportunity to improve their rank. For rhesus macaque males, low dominance rank may be a transient condition associated with a particular stage of life. Males may change ranks many times in their lifetimes, and whether or not they make it all the way to the top depends only on themselves. Females are just as opportunistic and eager to climb the hierarchy as males are, but because of the constraints of the matrilineal social structure, the Cinderella dream of becoming the alpha female rarely comes true. In the rhesus world, miracles don't happen, but just as there are in our world, there are occasionally individuals who take power into their own hands and start revolutions that will change their lives, and the lives of those around them, forever.

Chapter 5
WARS AND REVOLUTIONS

Xenophobia

People have a natural tendency to like others who are similar to themselves and dislike those who look or act different. Psychologists call this phenomenon "ingroup" versus "outgroup" behavior.[1] Racism is an example of outgroup behavior. We assign other people to groups all the time, and many times these groups exist only in our minds. Sometimes objective differences between groups do exist, but they are so small as to be meaningless. Biologists classify animals into different species on the basis of genetic differences or differences in their morphology. Using a biologist's standards, the concept of race doesn't make a lot of sense because, aside from the color of their skin and a few other minor characteristics, people from different races are biologically very similar and unquestionably belong to the same species. Unfortunately, when the human mind has the predisposition to work in a certain way, facts don't seem to matter that much. Acceptance of people who look or act different doesn't come naturally; we have to be educated about it, but with education, we can even develop a taste for it. The reason racism is so difficult to eradicate, however, is that differences between the races, though meaningless, are always there, right in our faces.

Our aversion to strangers is not simply a case of fear of novelty, or what we call neophobia. It's something more specific that applies only to other people, and therefore it deserves a term of its own. That term is xenophobia. To know what xenophobia is and what it does, we need only look at human history and see what has happened every time people from different continents have come into contact with one another. The first reaction has always been to try to exterminate or enslave the others. Even among people who live on the same continent, there is always the risk that some people's tendency to classify others as ingroup or outgroup will get out of hand and escalate into a war. When Australians first landed in the middle of a village in New Guinea in a helicopter—something that no one in the village had ever seen before—and offered the village chief a ride, the chief enthusiastically accepted, but immediately picked up a large rock. When asked what the rock was for, he answered that he was going to drop it on the heads of his tribe's neighbors.[2] This is a clear example of how much more powerful our xenophobia is than our fear of novelty.

Much like people, rhesus macaques don't like strangers, and their first responses to them are fear and aggression. When rhesus macaques see their image in a mirror for the first time, what they see is a rhesus macaque they've never seen before. The stranger in the mirror is staring at them, and they react to it the way Robert De Niro does in the film *Taxi Driver,* by screaming at the mirror, "You talking to me? You talking to *me?*" They fly into a rage and try to hit and bite their own image. When chimpanzees see themselves in a mirror, they are puzzled at first, but most of them eventually learn that they are looking at their own image and start using the mirror to pick their hair or clean their teeth. Researchers have been putting rhesus macaques in front of mirrors for years, but so far no monkey has shown clear signs of self-recognition.[3] One possible explanation for this difference is that rhesus macaques are not as smart as chimpanzees and don't have the cognitive skills necessary for self-recognition. Another explanation is that rhesus macaques are simply

FIGURE 8 Adult female threatening the observer. (Photo: Stephen Ross.)

too xenophobic and aggressive to be able to figure out that the mirror reflects their own image. In other words, rhesus macaques have such a strong tendency to hate strangers and to respond to threats from strangers with fear or aggression that they can't suppress those impulses when confronted by their own image. It's not that they are not smart enough to learn that they are looking at their own image in the mirror. It's that they are too xenophobic and aggressive to give themselves the chance to learn what the mirror is and what it does.

When it comes to strangers, the only exception to xenophobia involves sex. Although people tend to date and marry members of their ingroup, men and women are often attracted to outgroup people and are quite happy to have casual sexual relationships with them. Something very similar happens in rhesus macaques as well. During the mating season, migrating males approach a group and try to lure the females away to have sex with them. If males tried to do this at the wrong time of the year, when infants are born, they would get a strong xenophobic response from everybody in the group, especially the females with babies. Six months later, however, the raging sex hormones in the females' bodies make something click in their brains, and they assume a very different attitude toward these males. All of a sudden, the females are happy to respond to the males' solicitations and run off, leaving their screaming infants behind, have sex with the males behind the bushes, and come back to the group minutes or hours later acting as if nothing had happened. In some cases, they like the experience so much that they encourage their lovers to follow them and join the group. Clearly, the process of male migration between groups has occurred for thousands, maybe millions, of years, and both males and females are well prepared to deal with it.

Unlike males, female rhesus macaques don't routinely transfer between groups. So if one day a crazy female decides to emigrate, or if female transfer is artificially done in captivity—that is, people take a female from a group and put her in another group—nobody has a biological predisposition to help that female, no matter what time of year it is. What you get is a xenophobic response in full blast. Things are made worse by the matrilineal social structure and dominance hierarchy of rhesus macaques, which are very resistant to change.

FIGURE 9 Adult female threatening another individual and holding up her tail to solicit support. (Photo: Dario Maestripieri.)

The only way a female enters rhesus society is through birth, and with the continued help and assistance of her family members. Any female that breaks the rules and tries to enter society in some other way will pay a high price. Chances are that she will be attacked and killed by other females.

Adult males are as xenophobic as females are, but if a new female entered their group in the middle of the mating season, they probably would not object to it. She would be another female they could have sex with. To the resident females, however, a new female is just a competitor, and unfortunately for her, she has no family or friends to help her. If the new female somehow manages to survive the aggression received upon her introduction into the group, she will enter the dominance hierarchy at the very bottom. The lowest-ranking females in the group will make sure that this is the case. The lowest-ranking females in the group, in fact, will be the individuals who immediately attack the newcomer and who continue to harass her for days, or weeks, or months, long after everybody else has gotten tired of it. This is because they finally have somebody they can dominate and finally get the chance to express all of the aggression and frustration they have repressed in a life spent at the bottom of the hierarchy. A female newcomer also gives the lowest-ranking females the chance to form coalitions with higher-ranking individuals and redirect aggression and, finally, the chance to have a scapegoat of their own. Introducing a new female into a group of rhesus macaques is a bad idea and shouldn't be done, but it's the best thing that could ever happen to the lowest-ranking females in that group.

Wars

Female rhesus macaques don't wander around on their own, so if one day a new female shows up near a group, there must be another group behind her. This can mean only one thing: war. To the mind of a rhesus macaque, an encounter with another group is probably like being hit by a tornado. The work of a lifetime spent classifying other group members as male or female, kin or nonkin, low ranking or high ranking, friend or foe, becomes instantly meaningless. All the cards each rhesus macaque has patiently piled up on the

FIGURE 10 Training to fight: Rough-and-tumble play among three male juveniles. (Photo: Dario Maestripieri.)

table day after day, year after year, are instantly reshuffled. Kinship and rank are forgotten. There is another deck of cards on the table, and it doesn't matter if that deck also has jacks, queens, and kings. The backs of the cards look different: ours are blue and theirs are red. When there is a real outgroup out there, there can be only one ingroup in here. It's US against THEM. Even a low-ranking rhesus loner with an ingroup of one, who hates everybody else around, becomes an instant patriot. Every drop of xenophobia in rhesus blood is transformed into fuel for the battle. The rage of the subordinates over their miserable lives finds its natural expression in their outgroup behavior. The group has never been so cohesive, and it turns itself into a real army, a war machine.

We don't know how rare or frequent intergroup encounters are in the wild because they haven't been studied much. All we know about them comes from the rhesus colony on Cayo Santiago. Every group on the island has met every other group at some time and fought with them. Groups have dominance relationships with one another, just as individuals do. Dominance works the same way for groups as it does for individuals, except that for groups there is a simple

FIGURE 11 Individuals of the same group foraging together on Cayo Santiago. (Photo: Dario Maestripieri.)

rule: large groups dominate small groups. The small groups try to stay out of the way of the dominants as much as possible. When food is dumped on the island, the top-ranking group eats first, and the others eat later. There is a fixed schedule of meals that the various groups have worked out with one another, so that each group has to wait until the group that ranks just above it is done eating. Unfortunately for them, the menu for the low-ranking groups is always the same: leftovers. The dominant groups also get to hang out in the most comfortable and scenic parts of the island; they walk on the roads and lie down on the grass to relax while admiring the landscape. The subordinates must climb rocks and cliffs and sleep in the bushes.

Despite the meal schedule arrangements and the travel routes that regulate monkey traffic on Cayo Santiago, traffic accidents do occur. When a subordinate group is five minutes early for lunch and the dominants haven't left the restaurant yet, the tension in the air is so thick you could cut it with a knife. The older subordinate monkeys who have seen this happen before try to stay calm and tell the others, "We aren't that hungry after all, our favorite food probably isn't ready yet, let's go have a drink and come back later." Too late. Murphy's law has an ancient evolutionary history, and rhesus macaques discovered it long before we did: anything that can go wrong will go wrong. Somebody is bound to do something stupid and cause a disaster. And so it happens. An infant from the subordinate group wants to play with an infant from the other group, so he walks into the food compound and immediately gets smacked by an adult female. A young and hot female from the dominant group thinks a subadult male from the other group looks cute and makes a sexual overture to him. The male's mother doesn't think her boy is ready for sex yet and threatens the female. All hell breaks loose. Two or three females from each group start screaming and fighting, and soon they recruit the rest of their groups. The two groups assume their combat positions. Adult males and females without infants are in the front lines, while females with young babies sit in the back, watching and screaming. There is good reason for them to stay behind. I once saw a female with a baby throw herself into a battle with another group, but in the midst of all the fighting her baby was yanked from her, tossed in the air, and stepped on by other monkeys. Females without babies fight in small groups, side by side with their mothers and sisters. Adult males, however, often bear the brunt of the confrontation because males and females from the other group may gang up against them. A lot of the fighting is just screaming and threatening, but when contact occurs, the biting and scratching can be serious. The fight can last a few minutes or hours. Eventually, the subordinate group slowly moves away from the dominants, or gets chased into the water by the dominant group. If you live in a subordinate group on Cayo Santiago, you'd better be a good swimmer.

To Kill or Not to Kill

On Cayo Santiago, casualties of wars between groups are rare, but things may be worse in the forests of India, where rhesus macaques are not just angry all the time, but also hungry. Wars between groups of rhesus macaques with lots of casualties haven't really been observed anywhere, but they probably happen. What makes the difference between wars with casualties and those without them is whether or not the groups have met and fought before; that is, whether or not they have already established a dominance relationship and how stable or unstable it is. When groups meet for the first time, the macaques seem to be willing to kill those from the other group, whereas casualties are rare or nonexistent if the groups have fought before and have a stable dominance relationship. One explanation for the difference between these situations is that xenophobic responses toward individuals that haven't been met before are a lot stronger than those toward more familiar individuals. Another kind of explanation, which is fully compatible with the previous one, has to do with the costs and the benefits of lethal aggression. The reasoning is the same whether it applies to fights between individuals or between groups.

Rhesus macaques live in groups so that they can protect themselves from predators and fight other groups for food and space. In the rhesus world, strength is in numbers. Dominants don't kill subordinates in their group because they need them. Subordinates are valuable commodities because they fight hard during battles with other groups and because they get eaten first when a tiger attacks. Dominants harass and intimidate their subordinates all the time, but they are careful not to cause serious injuries or death; otherwise, the dominants would pay a price for their actions. Losing somebody you need is one of the costs of lethal aggression. As long as the dominants keep their power and dominance relationships are clear, the subordinates can and must be kept alive and well so that they can do all the dirty work for the dominants.

Dominant groups don't really need subordinate groups, but as long as it's clear who's dominant and who's subordinate, there is no point in trying to exterminate other groups. Rhesus macaques haven't invented weapons of mass destruction yet—or at least we haven't found them yet—so if the monkeys of dominant groups tried to kill all the monkeys of subordinate groups, they would risk being injured in the process. Risk of serious injury or death is another cost of lethal aggression. If groups fight over food, no single meal is worth the risk of trying to kill your fellow macaques. Dominance can deliver the same meal without getting your hands dirty. All that is needed is some aggression to maintain dominance and the status

FIGURE 12 How a war begins: Two adult females confront a female from another group. (Photo: Dario Maestripieri.)

quo, between groups as much as within a group. Essentially, dominance is a trick invented by powerful individuals to reap the benefits of aggression without paying its costs, or while paying as little as possible.

Dominance relationships between groups on Cayo Santiago are well established and stable, and the traffic accidents on the island rarely change the status quo. Wars between groups on Cayo Santiago rarely have casualties simply because, these days, killing another monkey, whether from the same or another group, is simply not worth it. Things were very different, however, when the rhesus macaques first arrived on Cayo Santiago. When hundreds of rhesus macaques captured in India were dumped on the island, there was a bloodbath. That day, killing other monkeys was well worth it.

When individuals or groups meet and fight for the first time, they are fighting not for a single meal, but for power, and power is not about this or that meal, but potentially every other meal for the rest of their lives. The stakes are much higher. The potential benefits of winning a fight go up dramatically, and so do the price the monkeys are willing to pay and the risks they are willing to take. Once again, it all comes down to economics, and the macaques are good accountants. When two groups have their first encounter, they must establish a dominance relationship, and because the structure of power is so stable in the rhesus world, establishing dominance has a huge and long-lasting value, so the monkeys should fight to the death. No one, to my knowledge, has observed the first fight between two groups of rhesus macaques that have never met before, but I would expect that fight to produce many casualties on both sides. The day the rhesus macaques were first dumped on Cayo Santiago, they all fought to establish dominance over everyone else, and killed others or died in the process.

It should be apparent by now that there are two kinds of aggression in the rhesus macaque world: aggression to establish dominance and aggression to maintain it. Aggression to establish dominance occurs between individuals or groups that haven't met before and can be lethal. Because there is so much at stake, rhesus macaques are willing to kill others to become dominant, or to die in the process. The other kind of aggression occurs between individuals or

FIGURE 13 Escalation of conflict: The lone female is supported by another female from her group. An adult male and a juvenile are getting ready to intervene. (Photo: Dario Maestripieri.)

groups that have met and fought before; it's aggression used by dominants to maintain and consolidate their power. Sometimes, in the course of this kind of aggression, subordinates may "protest" against the dominants or even fight back a little, and mild squabbles ensue. This may happen when the fight is about a meal and the subordinates are very hungry, or when a subordinate group wants to protect one of their infants from another group. However, because the fight is about a single event or individual, it's not worth fighting to kill or risk being killed.

It would be worth it for subordinates to fight to the death only if the fight wasn't about something in particular—an object, or an individual, or a situation—but was instead about changing their dominance status once and for all. This would be a third kind of aggression: aggression to subvert dominance. This aggression should be as lethal as aggression to establish dominance because the stakes would be just as high. This kind of lethal aggression is not between strangers that have never met before and don't have a dominance relationship yet, but between individuals or groups of individuals that have met before and already have a dominance relationship. The fight wouldn't be about establishing dominance, but about changing it. It wouldn't be a war, but a revolution.

Revolutions

One puzzling question about the structure of rhesus macaque society, as well as any human society with a class structure in which low-ranking individuals are oppressed and exploited by the dominant class, is "Why don't the subordinates just leave and go somewhere else"? In rhesus society, males leave their group as a rule, and being subordinate probably facilitates the process of male emigration around the time of puberty or secondary transfers between groups later in life. But why don't subordinate females leave their group too and look for a better life somewhere else? Low-ranking females benefit from living in groups just as high-ranking females do, and groups need to be of a certain size to be effective in protecting against predators or fighting over food with other groups. Subordinate females would get into a lot of trouble if they left their group on their own or in family groups too small to protect themselves or compete effectively with other groups. When groups grow beyond a certain size, so that splitting the group in half would result in two groups large enough to make it on their own (usually over a hundred individuals), such splits do occur. After such group fission, the matrilines in the top half of the hierarchy go one way and form their own group, and the matrilines in the bottom half of the hierarchy go a different way and form another group.[4] Half of the monkeys that were middle ranking before the split won't benefit from it because they will be at the bottom of their new group, while the other half of the middle-ranking monkeys will become the top-ranking matriline in the other group.

In groups that live in captive enclosures, the subordinates have no option but to stay where they are and put up with harassment by the dominants. Even their patience, however, has its limits. When those limits are reached, the result is the French Revolution, or, in the jargon of primatologists, a "matriline overthrow."[5] During these revolutions, members of one or more matrilines viciously attack the members of one or more higher-ranking matrilines, and after a bloody battle, they overthrow and eventually outrank them. The dominants don't get their heads chopped off like the king of France and his court, but they may be killed in some other equally violent way. The other main difference between rhesus matriline overthrows and the French Revolution is that in rhesus macaques, revolutions are entirely a female business. They are initiated by females against other females and result in changes in the structure of female power within a group. Adult males typically are not involved in these revolutions and are not affected by them, unless they are members of the matrilines involved in the overthrow (that is, they are young and have not yet emigrated), or unless they are crazy enough to get involved in the fighting when war between matrilines breaks out.

Matriline overthrows begin when members of one or more matrilines suddenly and concertedly attack the members of one or more higher-ranking matrilines. In many cases, the matriline being attacked is the top-ranking matriline in the group, or the top and the second-ranking matrilines together, while the attackers are the middle-ranking matriline or the matrilines that rank just below them. Fierce fighting can last a few hours, or in rare cases, several days, and it ends when the members of the matriline being attacked cease fighting back, either because they are dead or because they accept the new situation and begin to show submissive behaviors to their aggressors. After a successful overthrow, the survivors of the overthrown matriline fall in their dominance rank below the matrilines that attacked them, and in some cases all the way down to the bottom of the hierarchy.

During their wars and revolutions, people commit cruel and atrocious acts of violence against other people, acts they couldn't even conceive of during their normal lives. Likewise, the behavior of rhesus macaque females during a matriline overthrow becomes dramatically different from anything they would do in the course of their everyday lives. Aggression and fighting happen every day in a rhesus macaque group, and only a small minority of fights result in serious injuries. Most aggression between the members of a group involves threats and chases without physical contact, or when hitting and biting occur, they are typically directed toward less vulnerable body parts, mostly the back. Students of animal behavior call this "ritualized" aggression because it involves relatively stereotyped patterns of behavior that do not cause serious injury. Rhesus macaques, however, can easily switch from this ritualized aggression to lethal attacks that can kill another monkey in a matter of a few seconds. When rhesus macaques attack to kill, they bite vulnerable body parts, such as the face and the genitalia, arms and legs, or fingers and toes—that is, body parts that are not protected by thick layers of fat or muscle—and these bites can result in massive bleeding and fatal or permanent damage. The fighting associated with a matriline overthrow typically ends with some dead bodies on the ground and many individuals covered with blood and missing various body parts.

The general reason why this fighting is so lethal has to do with the costs and benefits of this form of aggression, and in particular, what's at stake in the fight. Rhesus macaques, however, obviously don't think about what's at stake or engage in mental calculations of the costs and benefits of their behavior. Instead, the fight escalates into serious violence because the individuals being attacked don't show the submissive signals that would normally terminate a fight. The members of the matriline(s) under attack are being attacked by individuals they have defeated in the past and who normally show fear and submission to them. If they experience an emotion upon being attacked, it is probably not fear, but surprise and outrage, and their immediate response is to fight back. However, the fighting associated with an overthrow is not just more intense or more prolonged, but also qualitatively different from the fighting of everyday life. Although it is always difficult to prove that the behavior of animals is motivated by specific intentions, rhesus macaque females fight as if they have an intention to kill. A similar difference in patterns of aggression is observed when chimpanzee males fight with other males of their own group, who are often their brothers, and when they fight and kill males from other groups. In one case they simply slap and push the others around, whereas in the other case, the aggressors bite the faces and the genitalia of their victims, or rip them off their victims' bodies.

Although there may be cases of attempted matriline overthrows that fail to achieve their objective, most known cases of overthrows are successful. This is only in part because failed attempts at overthrows may go unnoticed by people. Rather, it is probably because full-blown matriline overthrows occur only when they are likely to succeed. An increase in the size of a group results in social instability and increased risk of matriline overthrows. Matriline overthrows occur within a group when the top-ranking matriline, or the top-ranking and the second-highest-ranking matrilines combined, decrease in size relative to some middle-ranking matrilines due to the deaths or artificial removal from the group of several animals within a relatively short time. Dominant matrilines are typically larger or have a greater number of adult females than lower-ranking matrilines, and therefore base their power on their capacity to recruit many effective combatants when fights erupt with other matrilines. A dominant matriline may also be weakened and challenged if its matriarch, the alpha female, dies or is sick, or if several of its members are too old or sick to fight effectively. Researchers working with large groups of rhesus macaques know that the removal of an alpha female or several members of her matriline from the group has a high risk of triggering a matriline overthrow.

But what exactly triggers an overthrow? Do members of middle- or low-ranking matrilines keep track of the size or demographic structure of their group, or the health of the dominant females, and quickly devise plans for a revolution when they perceive the conditions to be opportune? Probably not. Although the death or disappearance of the alpha female is unlikely to go unnoticed by other group members, it is unlikely that this is consciously seen as an opportunity for an overthrow. More generally, it is unlikely that a matriline overthrow implies any rational or conscious planning by the macaque revolutionaries. Rhesus macaques are preprogrammed to take advantage of opportunities for social advancement when they arise, without wasting too much time on conscious thinking. Potential opportunities for overthrows probably arise on a daily basis in the life of a rhesus macaque group, because every day there are quarrels between members of one matriline, especially youngsters, and those of another matriline, and those quarrels have the potential to escalate into huge battles involving all the adult females of the two matrilines. Normally, however, when the dominant matriline has a strong grip on power, this escalation does not occur, because the alpha female and her family members are quick to support their relatives and effectively terminate the fight. If support does not arrive, however, or is slow and ineffective, the quarrel between the two youngsters will escalate into a large fight between their families, one in which the members of the lower-ranking matriline gain confidence in their ability to win. As in the case of Buddy, the adolescent male who was reintroduced into his group when he had not yet fully recovered from anesthesia, a display of weakness or lack of appropriate agonistic responses in one or more individuals will elicit vicious aggression by others. Therefore, the failure of members of the top-ranking matriline to adequately protect their youngsters or any member of their family on any given day could potentially trigger a revolution that will result in their overthrow and, in some cases, their deaths.

Whether the overthrown individuals will be killed or survive and be tolerated by their group members depends in large part on their willingness to accept the new situation and show submissive behavior to the new ruling family. Some individuals adjust their behavior relatively quickly, whereas others continue to fight back until they are killed. These different responses seem to depend more on the personality of the individuals than on their previous dominance rank or age. For example, there are known cases of alpha females who were overthrown but fought back until their deaths, and cases of alpha females who accepted the power reversal and continued to live in the same group for many years after the overthrow, but as the lowest-ranking female in the group.

Surviving a successful overthrow requires a dramatic change in the behavior of a formerly dominant female. This female may have spent her entire life at the top of the hierarchy and bullied every other individual in her group. All of a sudden, she has reason to be afraid of everyone in the group and must tolerate being bullied by individuals who until the day before were her favorite victims. These bullied individuals, of course, will do their best to make this task as difficult as possible, as they finally have the opportunity to take revenge for years of harassment and humiliation. This explains why, when an alpha female is overthrown, she is likely to fall to the very bottom of the hierarchy and be outranked even by the surviving members of her own family who were overthrown along with her. Everybody in the group has an ax to grind with an alpha female who loses her power, and the burden of this aggravation is such that the female will fall to the bottom of the hierarchy and will never be able to rise again.

Chapter 6
SEX AND BUSINESS

Biological Origins of the Sex Industry

In the United States, sex is a multibillion-dollar industry. Sex sells. Selling and buying sex itself is called prostitution, pornography, and many other names. Using sex to sell anything else—food and clothes, magazines and books, TV news and Hollywood movies—is called good business. Almost anything sells a lot better if there is sex attached to it. Men are big buyers of sex: they start at puberty and don't stop until they drop dead. Business executives in the advertising and entertainment industries seem to understand what sex means to people a lot better than the psychologists, sociologists, and biologists who spend their careers studying sexual behavior. Businesspeople understand that sex is a valuable commodity and that many people are willing to pay a lot for it. It takes a lot of advertising and brainwashing to convince people to buy one brand of cereal instead of another because it tastes better or because it's better for your health, especially if the two brands taste exactly the same and are both equally useless for your health. However, if you put the fantasy in a man's brain that he can have sex with an attractive woman if he eats a particular brand of cereal, there is no need for aggressive and expensive advertising campaigns because men are already primed and eager to buy the sex cereal.

People are not the only animals that use sex for business purposes. Other primates do it too, and as you might guess, rhesus macaques can be quite Machiavellian about it. The key to understanding how sex has turned into a business is the nature of primate female sexuality. Many primate females have menstrual cycles very similar to those of women. Although they are fertile only when they ovulate—a few days around mid-cycle—they can have sex on any day of their cycle. They differ in this way from many other animals, in which females have sex only during their fertile period, called estrus. In these animals, females simply can't or don't want to have sex at any other time. In female guinea pigs, for example, the vagina is sealed by a membrane when they are not in estrus, and they couldn't have sex even if they wanted to.[1] When female guinea pigs are in estrus, their sex hormones induce chemical changes in their bodies that dissolve the vaginal membrane. The same sex hormones also act on the females' brains and turn on their desire for sex, but only for those few days of fertility. In the guinea pig and in many other animals, sex is really about making babies, and there aren't many opportunities for business. In rhesus macaques, people, and some other primates, female sexuality works differently. As my former postdoctoral advisor, Kim Wallen, says all the time, in primates sex hormones don't affect whether females can or cannot have sex, but only whether they want to or don't want to.[2] Female primates have much greater control over their own sexuality.

Back in the 1920s, a British biologist named Solly Zuckerman noticed that monkeys seemed to have sex all the time. In particular, he noticed that even though monkey females come into estrus in the same manner as other animals, they seem able and willing to have sex outside their fertile period as well. So Mr. Zuckerman thought that sex was the glue that kept primate societies together and that primate social life revolved around sex.[3] We now know that sex is not the only reason primates live in groups. Group living has many benefits, including cooperation to find food and protection from predators. There are also forces other than sex—kinship, for example—that keep these groups together. Nevertheless, sex is an integral part of primate social life.

Early laboratory studies of rhesus macaque sexual behavior seemed to confirm Zuckerman's observations. Researchers put a rhesus female and male in a small cage and observed that they had sex every day of the month.[4] The researchers who did these studies happened to be men, so their biased male minds reasoned that rhesus males and females had sex every day because that's what the male wanted and the female was just passive and going along with it. They also thought that the female's main role in sex was to be attractive, and that rhesus females were more attractive to the males when they were fertile because they smelled nicer.

Now, it is probably true that the rhesus males in the small cages wanted to have sex every day. The rhesus females who had the bad luck of being used as subjects in these experiments, however, were not being passive at all. They were doing business with the males, exchanging their sexual availability for their safety and survival. If you are a rhesus female and you find yourself trapped in a small cage with a large and aggressive male, you soon realize that saying "Not today, dear; I'm just not in the mood" is not an option. In that small cage, there was no place to run, no place to hide, and no family members around to help. There wasn't even room to turn around, smile at the male, and say "No thanks." The female had to have sex every day with the male to stay alive. As for the part about female attractiveness and smell, the rhesus female in the cage wasn't the only one doing business. Some of the researchers who did many of these experiments claimed to have discovered a sex pheromone, a chemical substance that made females smell very attractive.[5] One of them tried to get a patent for it and sell it.

When Clarence Ray Carpenter observed rhesus macaques having sex on Cayo Santiago in the 1940s, he saw something very different.[6] Rhesus females had sex with males mostly at mid-cycle, and they were not passive at all; they were in charge. Many other researchers who studied sexual behavior in large rhesus groups saw the same thing.[7] Rhesus females have many ways to let males know they are interested in sex and want to mate. They follow the males around, sit

FIGURE 14 Mating time: A foot-clasp mount. (Photo: Stephen Ross.)

next to them for hours, and every now and then stand up, raise their tails, and stick their butts right into the males' faces. If the males don't pay attention, the females slap them and smack them, and in case the males have any doubts as to what the females have in mind, the females will jump on their backs and start mounting them to remind them how it's done. Rhesus females also have many ways to let males know they're not interested. Most of the time, they just ignore them and walk away from them. If the males get too persistent, they may threaten them and scream for help from their family members. In a large group where rhesus females are surrounded by and helped by their family members, it would be a very bad idea for a male to get rough with a female if his attention is unwanted. Male sexual coercion doesn't exist in the matrilineal and female-dominated rhesus macaque society.

Rhesus females are very interested in sex on the fertile days of their cycle, but also attend to the business of sex every other day of the month. To understand how this works and how similar or different things are in rhesus and human societies, we need to look at sex from both the male and the female perspective.

The Male Perspective

Males have sex for the same evolutionary reasons females do: to produce surviving children and grandchildren and to make sure that copies of their genes are passed on to future generations. Clearly rhesus macaques don't know that there is a connection between having sex one day and the birth of an infant six months later. They want to have sex because it's fun. Human males are supposed to know that nine months after they have sex with a woman, she might give birth to their child. Some human males, however, just don't seem to care.

Among rhesus macaques, males don't help with parenting at all. They don't know who their offspring are, they don't know that they have offspring, and they don't even have a concept of offspring or fatherhood.[8] Every now and then they see a baby pop out of a female and have no idea how that happened or that they had anything to do with it. Because they don't help with the kids, rhesus males are preprogrammed to produce as many of them as possible, and hence have sex with as many females as possible, as often as possible.

In chimpanzees and other great apes, males don't help raise their offspring either, so it's likely that human males started getting involved in the parenting business only very recently in their evolutionary history. The brains and hormones of human males are still struggling to adjust to this new situation, and in some cases, it looks as if they haven't adjusted at all. When it comes to sex, the human male brain still works a lot like the brain of a rhesus macaque. The voice in men's brains that tells them to have sex with as many females as possible is still very loud; it's been there for a long time and is not going away anytime soon. The voice that tells them to be good and responsible fathers is young and soft, and some men simply choose not to listen to it.

When it comes to sex, rhesus and human males are a lot more similar than they are different, and explaining their behavior doesn't really require a Ph.D. Males just want to have sex. Sperm is so cheap to make, and males have so much of it all the time, that they are just happy to give it to anybody, any place, any time. Theoretically, every male would be happy to have sex with the whole female world population, do it in a couple of weeks or whatever it takes, and then start over again.[9] Luckily for both males and females, this is not an option. Males are lucky if they can find a single female who is willing to have sex with them. And it's not really a matter of luck. Unless you look like Brad Pitt, have Donald Trump's money, or are the editor of *Playboy* magazine like Hugh Hefner (or you are their rhesus equivalents), there is a lot of work involved.

There are two kinds of things that get in the way of a male's plan to have sex with all the females on the planet. The first big problem is that all the other males have exactly the same idea, and therefore there is competition. The second problem is that a male would have to convince every single female in the population, one at a time, to have sex with him and not with all the other guys who are knocking on her door at that very same moment. These two problems put a lot of pressure on a male. Evolutionary biologists call this pressure sexual selection, and the person who first described it was Charles Darwin.[10] In animals such as deer and some birds, fighting off other males and attracting females is one and the same. Females watch the males fight with one another and then simply have sex with the winner.[11] In rhesus macaques and humans, things are slightly more complex.

Rhesus males become sexual competitors the day they have their first sexual thought and act on it. When males reach puberty, at four or five years of age, sex ceases to be a form of play and becomes the complex interaction we're all familiar with, but without the human penchant for candlelight, soft music, and sexy lingerie. A young rhesus male's discovery of his sexuality doesn't go unnoticed, and isn't welcomed, by the adult males in the group. From the day a young male has real sex with a female, he will be harassed and attacked by the older males, who will eventually kick him out of the group. Emigrating to another group, however, doesn't automatically solve the young male's problem. Although rhesus females may be happy to welcome a new male into their group—if it's the right time of the year—the resident males don't exactly roll out the red carpet. This means that in order to have sex, the young male will have to fight. How much fighting and how tough it is depends on the male's fighting abilities and his willingness to take risks. In theory, the migrating male could use his martial arts and fight the group's alpha male and all the other males at the same time, like Keanu Reaves does with the bad guy and his clones in *The Matrix*. If he's The One, he will win, become the new king, have sex with all the females in the group, and live happily ever after. In practice, more often than not, the migrating male will be fearful and submissive, enter the new group at the bottom of the hierarchy, and try to work his way up one step at a time.

From the day the new male is officially part of the new group, he has to work on two sets of problems: how to climb the male dominance hierarchy and become the alpha male, and how to be nice to the females and gain their tolerance and sexual favors. This is the point in a male's life where his Machiavellian intelligence becomes critical. If the male plays his cards well, he will get to the top of the hierarchy in a few years and manage to sire a few offspring in the process.[12] As the new alpha male, he will have uncontested sexual access to all the females in the group (assuming that the females like him, of course). The longer he remains the king, the more offspring he will produce. In the meantime, however, the Machiavellian plots and schemes of all the other alpha male wannabes will be directed at him more and more frequently. Eventually one of wannabes will hit the jackpot. At that point, the deposed king's reproductive career could be over. On Cayo Santiago, some males who lose their alpha status manage to transfer to another group and reproduce again. Others end their careers being tolerated as low-ranking males within their own group, but with little or no chance for reproduction. In the forests of India, the deposed king would probably leave his kingdom, wander around on his own for a while, and die shortly afterward due to predation, sickness, or murder by other monkeys.

An alpha male must control the ambitions and sexual urges of the other males and be nice to the females. There seem to be two kinds of alpha males in the rhesus world, which probably reflect two basic personality types, or using the jargon of evolutionary biologists, two alternative mating tactics.[13] Alpha males of the first kind are control freaks who become obsessed with trying to prevent other males from mating. During the mating season, they become the sex police and watch all the females and males in their group day and night. If they see a female and a male engage in foreplay—that is, they start grooming each other—or if they catch a couple in the act, they will immediately rush to the scene of the crime and chase the female away (for some reason, alpha males always get mad at the female and never at the male). The problem with this kind of paranoid alpha male is that he stresses himself out and hardly gets to reap the rewards of his social status. At the peak of the mating season, there is so much sex in the air that trying to suppress it all is a full-time job with little chance of success. Alpha males of the other kind concentrate on having as much sex as possible without caring too much about what everybody else is up to. Clearly they are not blind or stupid, and they notice what's going on around them. You can see them clenching their teeth when they see or suspect something, but they won't intervene and chase the culprits. Instead, they just concentrate on their mission of impregnating all the females in their group as quickly as possible. It's like an assembly line; as soon as they are done with one female, they just move on to the next in line.

We don't know what makes a male type I or type II. I was once watching two groups of rhesus macaques at the Yerkes National Primate Research Center, and one group had the paranoid kind of alpha male while the other group had the assembly line kind. We also don't know whether the two kinds of alpha males end up siring a similar number of infants in their groups. All we know is that whichever way they choose to play the game, the mating season can be a very stressful time for an alpha male, and for males in general. On Cayo Santiago, adult males lose up to 10 percent of their body weight during the mating season and spend the following six months, while the infants are being born and raised, relaxing, fattening up, and getting ready for the next round.[14]

An alpha male mates with many females in his group, but especially with the alpha female and her family members. High-ranking females may be in better shape and more physically attractive (maybe they smell nicer too!), or maybe they are just around the alpha male more often. Rhesus males seem to like middle-aged females at the peak of their fertility and are not particularly attracted to adolescent or old females. High-ranking and middle-aged females may be not only the most fertile, but also the most likely to raise their offspring successfully, although it's unlikely that males are aware of that. Low-ranking females mate less because the dominant females try to keep them from doing it. During the mating season there is aggression not only between the males, but between the females as well. When high-ranking females see that subordinate females are consorting with males, they harass them and chase them away.

All males clearly prefer to mate with fertile females. If there is only one estrous female in the group, however, chances are she will be monopolized by the alpha male. Low-ranking males mate with any female they can. This often means that they end up mating with the unpopular adolescent females, with old and ugly females from low-ranking matrilines, or with females who are having their period. Low-ranking males try to hide from the alpha male when they mate. I once saw a low-ranking male hide and mate with a female behind a concrete culvert while constantly checking to see if the alpha male was looking. He ejaculated in two seconds and then walked away from the female, acting as if nothing had happened. Dominant males sometimes take many repeated mounts to be able to ejaculate, but the subordinate males' anxiety and fear of getting caught makes them ejaculate almost immediately.[15] This problem of anxiety and early ejaculation is shared by some human males. Rhesus macaques show us why the problem may exist, and that it's not really a problem. If you are a subordinate male, premature ejaculation may be the best way, or the only way, for you to inseminate females.

Males can tell when females are fertile mostly from changes in their behavior, but also by the reddening of the skin around the females' genitalia and in their faces. The physical signs of estrus in rhesus females, however, are not as clear as those in other primates such as baboons or chimpanzees, in which the skin around the females' genitalia swells up and assumes gigantic proportions. Rhesus males therefore cannot be absolutely certain of when females are ovulating, but must take educated guesses and trust what the females are telling them with their behavior. Females, for their part, have their own reasons for not being completely honest with the males. Females use males not only as providers of sperm, but for other services as well. For instance, females might decide that they want to mate with the alpha male when they are fertile and with other males when they are not so fertile, for Machiavellian reasons that will be explained later.

For males, trying to find fertile females who are willing to have sex takes a lot of time and effort, and the results aren't always encouraging. Rhesus and human males go through long periods in which they can't find any females who are willing to have sex with them. In rhesus macaques, that's because for six months of the year females are interested in babies and have no interest in sex at all. In humans, there may be many different reasons why some men can't find a woman for months or years, but let's just blame it on bad luck. These large fluctuations in the availability of sexual partners are probably nothing new and have been happening to both rhesus and human males for thousands, maybe millions, of years, so that even their brains and bodies have begun to take notice. Rhesus males typically experience a big drop in their testosterone levels during the six months of the birth season and lose their libido.[16] What's the point of being sexually aroused all the time if the females are just not interested? My research collaborators and I once gave some rhesus females injections of a sex hormone, estradiol, during the birth season. All of a sudden these females became interested in sex and started following the males around and pestering them with their sexual invitations.[17] The males, however, were just not interested. In rhesus macaques, male sexual motivation has become seasonal, just like that of females. This is not a matter of conscious choice, however. The brains of rhesus males have hardwired biological clocks that can tell what season it is from the amount of daylight and the temperature. Depending on the season, their brains send messages to their testicles and tell them how much testosterone they need to produce. This, in turn, tells the male how sexually motivated he needs to be.

People don't have sex only between spring and summer like the rhesus macaques on Cayo Santiago, but I suspect that male brains and bodies have figured out a way to keep track of temporal fluctuations in the availability of females and opportunities for sexual activity. We know that male sexual motivation is regulated by testosterone, and I suspect (but don't know of any strong data that prove it) that testosterone production is regulated by experience through a mechanism called positive feedback. Positive feedback means that

FIGURE 15 When females are not available or not enough: An adult male masturbating. (Photo: Dario Maestripieri.)

the more you have of something, the more you want, and the less you have, the less you want. I suspect that when a male has sex, or thinks he has the chance of having sex, his body produces more testosterone, and testosterone increases both his sexual motivation and his sperm production. If a man, instead, goes through a period of bad luck with no sexual activity, and for some reason he doesn't think things are going to get better anytime soon, I suspect that his body produces less testosterone, and the man thinks less about sex, masturbates less, and produces less sperm as well. We know that stress and depression can reduce a male's sexual motivation, and I would guess that men who don't get any sex for months or years at a time aren't too happy about it. So, even though rhesus and human males have the potential ability to have sex all the time, in practice their sexual motivation probably goes up and down in relation to what goes on around them, the presence of sexually available females, and their experience with females.

Finding a female who is willing to have sex with him is a big problem in a male's life, but not the only one. Survival and comfort are also important. Males also like to have friends—male or female. For the most part, however, males would be happy if their sex life didn't interfere with their social life. Men would be happy if they could get all the sex they needed in the form of one-night stands—go to the bar in the evening, meet a woman, have sex with her, and get back home in time to watch the football game on TV. The more males try to keep the sexual and the social spheres of their lives separate, however, the more females try to bring them back together. Many times, females simply won't have the sexual without the social. In order to get sex from a woman, a man is going to have to spend time and money on her. That's because for males, sex is just sex, but for females, sex is also business.

The Female Perspective: Sleeping with a Stranger

Rhesus females are attracted to males they've never seen before, like the guys who hang out at the periphery of their group and try to lure them away to have sex behind the bushes.[18] In many cases females have sex with these males and never see them again. In a few cases, though, females will help these males join their group, work their

FIGURE 16 It's the mating season also for this tree on Cayo Santiago. (Photo: Dario Maestripieri.)

way up the hierarchy, and eventually displace the alpha male. The same rhesus females who have sex with strangers behind the bushes also have sex with the resident males within their group many times during the mating season and over several years in a row. So it seems that rhesus females like two different kinds of sex, casual sex with strangers and sex with familiar males with whom they have stable, long-term relationships. They are not the only primate females with these kinds of preferences.

Under certain circumstances women, too, like to have sex with a guy they've just met and will never see again. If men have one-night stands, this must mean that there are women out there who enjoy having one-night stands too. I really doubt that there are only a handful of women who enjoy casual sex and that all the men who have one-night stands do it with the same few women. Although women may choose, for a variety of reasons, not to advertise that they enjoy casual sex with strangers, I suspect that there are more women who like it than men think. Women also clearly like to have sex in the context of a stable relationship with a man they love, and that's something they advertise a lot.

The simplest explanation for the sexual preferences of rhesus and human females is that they have sex for two different reasons. Since males have a one-track mind, the notion that females may want to have sex for two different reasons is very confusing to them. Some men take a long time to figure it out, and others never quite get it. As an adolescent, I started out thinking that women were one-track-minded just like men, but that their track was different from ours: that they liked sex only when it was wrapped inside the love&relationship package. So, back in high school, when I had a crush on a girl and wanted to have sex with her, I offered her my love, friendship, and long-term commitment. Unless the girl found me totally repulsive, she entered an intense relationship with me with a lot of talking, holding hands, and attention on my part. In my mind, this was all going to end up in a lot of sex one day, but, of course, it never did. That same girl who spent whole days holding hands with me was having sex in the evenings with a guy twice my size, with a symmetrical face and a large jaw. And there may have been more than one.

Wait a minute. I just told you that women like to keep sex and social relationships together, but now I am telling you that this girlfriend of mine wanted to have a sexless social relationship with me during the day but casual sex with other guys in the evening. Am I contradicting myself? Plus, what kind of immoral girls did I know, who were cheating on me like this? Maybe deep down in the backs of their minds, females like to keep sex separate from their social relationships just like guys do. After all, this is the way things were at the very beginning, before humans and many of the other primates evolved. Sex was only about making babies and had nothing to do with anything else. Now things are different, but there could still be some trace of those ancient times in the female brain. The problem is that females would have a lot of trouble getting males to invest in a long-term relationship with just the promise of sex. A naïve young boy such as myself may have been willing to embark on these platonic but time-, energy-, and money-consuming long-term relationships with the hope that one day sex would fall from the sky, but I don't know many experienced boys or men who would be willing to do it. Women have to use sex to keep men around or else the relationship just won't work.

So, once women have made up their minds about who they want to have a long-term relationship with, they will have sex with that man and make him feel that they are doing something very special and unique with him. And by the way, my girlfriend was not cheating on me, but was quite honest about what she was doing. First, although she spent a lot of time talking and holding hands with me, she never told me she really considered me her boyfriend. Second, she made no secret of the fact that she was having sex with another guy in the evening. She told me that it was just sex and didn't mean anything to her, and that she didn't really care about this other guy, or not as much as she cared about me. She was telling me the truth. It wasn't until later—much later—in my life, that I discovered the two tracks in females' minds and started making some sense of their behavior. To this day, though, there are times when I can't tell which track a woman is on because women switch back and forth between the two tracks without letting men know what they are up to. Rhesus males sometimes face the same problems men do. They spend a lot of time making friends with particular females, grooming them and protecting them, but when the mating season finally starts, these females have no interest in having sex with them. Instead, they prefer to have sex with other males who are not their friends, or with males they don't know at all. So what kind of game are the females playing?

When rhesus females have sex with strangers behind the bushes, they do it for a couple of reasons. They want to be fertilized by unrelated males who might have good genes for their offspring, and they want to screen males as potential replacements for the resident males in their group. Although it's unlikely that the resident males were actually born in that group, when rhesus females mate with them, there is always a chance they could be mating with their fathers or brothers, or even their sons.[19] That would be very bad for the females and their offspring, for reasons explained in chapter 3. Mating with strangers ensures that inbreeding and its negative genetic consequences are avoided. But there is more to it than that. When rhesus males emigrate from their natal group, they have to survive a lot of adversity, and some of them probably end up starving to death or being eaten by predators. Others may waste a lot of time in all-male bands and just don't have the guts to show up near another group, lure the females away and have sex with them, and then be ready to fight off the resident males if they get caught. The males who make it all the way to knock on the females' door probably have pretty good genes, and by mating with them, females give their offspring a chance to inherit those good genes. If the males have genes that make them especially attractive to females, the females hope that their sons will have the same genes and father many offspring in turn. In evolutionary biology this is called the "sexy son hypothesis" of mate choice.[20] Of course, rhesus females are not conscious of any of this. They just have a preference hardwired in their brains for unfamiliar males with particular characteristics (for example, big and strong, with self-confident behavior), and this preference becomes especially strong after they have hung out with the same males within their group for a number of years. Chances are, of course, that the alpha male within their group started out his career like one of these guys and therefore has good genes as well. But the females have already mated with him, and he's probably the father of several of their offspring already. For a female, it's simply not wise to put all of her eggs in one basket. Females love shopping around and putting their eggs in many different baskets.

If rhesus females have sex with strangers in order to be fertilized by males with good genes, it would make sense for them to be interested in these guys only when they are fertile. In fact, females run out of their group to have sex with strangers only when they are in estrus and at the peak of their fertility. Otherwise, they don't bother. Some recent studies by evolutionary psychologists suggest that something similar may be going on with women as well. Women appear to be most interested in casual sex when they are at mid-cycle. At this time, women are especially attracted to strangers who are likely to have good genes; for example, guys with a very symmetrical face or a large jaw.[21] In men, these physical characteristics are signs of healthy growth and high testosterone levels (just think about Superman's face). If women are already in a long-term relationship and have an extramarital affair with one of these symmetrical and large-jawed guys, they may also be considering the possibility of replacing their current partner with the new guy. Like rhesus females, they are shopping around. But if their current partner is a good father and maybe also cooks, does the dishes, and takes out the trash, they will stick with him and get his help to raise the large-jawed baby boy that is already on the way.

So this may explain why rhesus and human females are interested in casual sex, when they want to do it, and who they want to do it with. But why aren't females *only* interested in casual sex? Why don't rhesus females have sex *only* with the strangers that knock on their door during the mating season? Why do both rhesus and human females also have sex with males with whom they have long-term relationships? The answer is that, in this case, what females really care about is not sex, but the long-term relationship itself. In this case, sex is just a means to an end, the way to keep the relationship going. This is the second track of their two-track minds, and this track is about business.

The Female Perspective: Relationships and Business

Before we explain why and how sex can be used for business, it may be useful to spend a few more words on the nature of female sexuality. Understanding female sexual behavior doesn't require a Ph.D. either, but some basic knowledge of biology and reproduction is helpful. Males produce sperm constantly, and in theory, they could inseminate a new female and produce a baby every minute of their lives. Females, however, have the potential to produce a baby on only a few days of the month, and if a baby is conceived, they can't produce another one for many months or years. For a very long stretch of our evolutionary history, sex actually had something to do with reproduction, so this difference in the reproductive biology of the two sexes raises the interesting, provocative, and politically incorrect idea that maybe, on average, females are not as interested in sex as males are. Let me hasten to qualify my statement and avoid any further anger from my female readers.

In rhesus macaques, estrous females are as interested in sex as males are and show it even more. When they are not in estrus, however, females seem to be less interested in sex than the average male. The physical pleasure a rhesus female derives from sex is probably just the same whether she's ovulating or menstruating. So it's not that females with their period don't enjoy sex, but they are just not as much in the mood for sex as when they are fertile. Their hormones and their brains have something to do with this. Something similar may be going on with women, although the changes in sexual motivation across the cycle are probably not as strong as in rhesus macaques. If you asked women to tell you when they have sex with their partners in the course of a month, they would probably answer that they have sex on weekends because that's when they get the chance to do it, especially if they and their partners have a full-time occupation. However, if you asked women to tell you when they have sexual desires or when they masturbate, that probably happens a lot more around mid-cycle than at any other time of the month.[22]

So it seems that while males' sexuality is hardwired in their brains and in their glands, female sexuality is equally hardwired at mid-cycle, when sex is tightly linked to reproduction, but not so hardwired at other times of the cycle. Both rhesus and human females have sex during their nonfertile periods, but the fact that they are generally less interested in it and less driven by their hormones means that they have more direct control over sex at those times—that they can choose whether or not they want to do it, or when, or with whom. This greater independence of their sexual motivation from hardwired mechanisms during nonfertile periods, coupled with the fact that sex is detached from reproduction and the fact that they generally care less about doing it, puts females in the perfect position to use sex for business purposes.

Sexual motivation is only part of the story. Basic knowledge of biology and reproduction also suggests that females should be a lot more careful about having sex and who they do it with than males are. In rhesus macaques, only mothers have milk to feed their babies, and if they aren't there to feed them, their babies will die. Ever since formula and bottle feeding were invented, women have had the potential ability to leave their babies with their fathers and move on to another man to repeat the pattern. The problem is that human brains can't keep up with the pace of technological progress. So mothers are, on average, much more reluctant and less likely to abandon their offspring than fathers are. This makes women a lot more careful than men about having sex and babies and about who they have them with. There is another important reason for female caution: women run a greater risk of getting hurt and catching infectious diseases through sexual intercourse than men.

Neither human nor rhesus males, as a rule, physically coerce females to have sex with them (although men have the tendency to use sexual coercion more than rhesus males do). This means that, because females are a lot more reluctant and choosy about sex than males are, sex between a male and female doesn't happen unless the female says it's okay. Even if I am all wrong about sex differences in sexual motivation and females want to have sex as much as and with as many partners as males do, the fact still remains that women are much more in control of sex than men are. Once again, we are talking about the average here. There are females, of course, who are not as in control of their sexual lives as they would like to be, and a few lucky guys like Hugh Hefner, who have far more control of sex than any other men on the planet.

Now, whenever someone controls something that someone else wants, this immediately creates an opportunity for business; that is, a chance to sell this thing or trade it for something else. But if it's true that females are in control of sex, what is it that human and rhesus females could possibly want from males that they could exchange for sex?

What Females Want

In the film *What Women Want,* the main male character played by Mel Gibson all of a sudden becomes capable of reading women's minds, finally figures out what women want, and gives it to them, thus becoming very popular with the ladies. Unfortunately, reading women's minds is not an option in the real world, and even if it were, I am not sure it would be sufficient to guarantee a man's success. Men would still have to look like Mel Gibson or have his money to be successful. As an alternative to reading women's minds, let's try to figure out what rhesus females want from their males and see if something similar might work for women and men.

One thing rhesus macaques and people have in common is that males are physically larger and stronger than females. Rhesus males also have sharp canine teeth that are almost twice as long as the females'. Human males have a much greater tendency for physical aggression and violence than females. Just look at who commits murders and fights wars in any human society. Males are physically dangerous, and females seem to be aware of that, so one thing females want is protection. They want a male to protect them from other males who might potentially harm them. Since being stronger and more aggressive can also help with defense against larger animals, females also want males to protect them from predators and other dangers. This is the case for rhesus macaques and was probably true for our hominid ancestors as well. Protection from other males and predators is probably the main reason rhesus females allow a few males to be permanent members of their female-bonded groups. Having sex with them is how they keep them around.

Human males don't have just physical strength and aggressiveness, but also political and economic power. Women want some of that too, and all the goodies that go along with it. So a woman must find a powerful man, keep him around longer than the few minutes necessary for one sexual encounter, and get him to invest everything he's got—time, energy, money, power, you name it—in her. The key to all this is social relationships. Social relationships are the vehicle with which females keep males around and obtain their help and their resources. Men need both sex and social relationships, but not necessarily with the same individuals or at the same time. When women have sex just for sex, they are just like men and just as happy to do it outside of relationships. However, when they use sex for business, they can't do it outside of relationships, or else it would just be prostitution, an unsophisticated and dangerous form of business that most women rightly avoid. So, as already mentioned in chapter 3, human females don't establish and maintain a long-term relationship with a male so that they can have sex repeatedly with him; instead, it's the other way around. Females consent to having sex repeatedly with the same male in order to establish and maintain a long-term relationship with him.

Some other things females want from long-term relationships with males are different for rhesus macaques and people. Since men help take care of the kids and rhesus males don't, one thing women want from them is time, energy, and money to help raise their kids. Help from a male in raising the kids has become so important in some human societies (but not in all of them) that in order to get it, women are even willing to give up their impulses to have sex with strange men. One condition set by men for helping with the kids, in fact, is the absolute certainty that they are the fathers of those children. This means that men expect complete faithfulness from their female partners.

Rhesus males don't make good fathers. Instead, if they have never mated with a particular female and she has a young infant, they may kill the infant. In many primates and other animals, males kill infants sired by other males so that their mothers will return to estrus more quickly and become available for mating.[23] Thus rhesus females need not only protection for themselves but also, and especially, protection for their offspring. The risk of male infanticide and the need for protection have led primate females to become very Machiavellian with sex. Rhesus females have two basic strategies to protect their infants from infanticide. One is to have sex with the alpha male in their group, especially when they are fertile. Male brains are probably preprogrammed so that if males have a lot of sex with a female and they are the only ones to do it (or at least they think they are), they won't feel like killing the baby that's born six months later. Not only that, but their brains are preprogrammed to increase the males' willingness to fight and protect the female and her baby from other males under these circumstances. The females, on their part, don't know any of this consciously, but their own brains may be preprogrammed to make them want to have sex with the males that can provide the most effective protection from infanticide.[24]

One strategy for protection against infanticide may not be enough. As we mentioned before, rhesus females, and primate females in general, don't like to put all of their eggs in one basket. What if they have sex only with the alpha male during the mating season, but the alpha male dies or loses all his power before the baby is born?

The Machiavellian rhesus females have a backup plan that involves double-crossing the alpha male. While they have a lot of sex with the alpha male and make him think he's going to be the father of their baby, they also have sex with all the other males in their group behind the alpha male's back, as well as with as many strange males as possible. Six months is a long time in the life of a rhesus macaque, and many things can happen. For example, in those six months, one of the other males in the group could challenge the alpha male and become the new king, or another guy could join the group and become the alpha male. So rhesus females hedge their bets and have sex with all of these guys to give each a chance of being the father of their baby. If any of these males are still around six months later, and especially if they have increased their power, they will be less inclined to kill the female's baby because they would risk killing their own offspring. Once again, none of this involves any conscious thinking or rational decision making on anybody's part. Males probably have a switch in their brains that turns their infanticidal tendencies on and off. Having sex with a female probably turns off infanticidal behavior toward that particular female's offspring for at least six months. We don't know this for sure about rhesus macaques, but something similar happens in male mice, who are even more infanticidal than monkeys. Male mice's infanticidal behavior gets turned off if they have mated in the window of time that would make it possible for any mouse pups they encounter to be their own offspring, and gets turned on again if they haven't mated for a longer time.[25]

So, by mating with as many of the males in her group as possible, a rhesus female gives all those males a chance at fatherhood and reduces the probability that they might want to kill her baby when it is born. If the alpha male, however, found out how Machiavellian the females in his group are, he would be quite upset. The more the females have sex with other males, the less likely he is to be the father of their babies, and the more he's going to want to kill those babies. Once again, rhesus females seem to know how to manage the situation, and they try to have sex with all the other guys without being seen by the alpha male. This way, all the males are happy, and the females obtain what they want from them.

As if this isn't Machiavellian enough, there is more to it. The alpha male in a group has already proved his skills, and possibly also the quality of his genes, by becoming an alpha male. The other guys in the group, however, haven't proved themselves yet. Some of them could be potential alpha males in the making and therefore have genes as good as, if not better than, the current alpha male's. On the other hand, some of the other males could be losers who will never become alpha males. Rhesus females improve their genetic odds by having sex with the alpha male when they are fertile and with the other guys when they are not so fertile. Until these guys prove themselves, the females will not entrust them with the fatherhood of their infants. The alpha male is happy this way, and the other guys are not too upset either. Some of the other males in the group, especially the very subordinate ones, are so happy to get any sex at all that they won't fuss about whether the females who do it with them are at the peak of their fertility. Second, females are not always a hundred percent honest in letting the males know how fertile or nonfertile they are. In rhesus macaques, ovulation is not advertised with giant neon signs, as in other species. In rhesus macaques, and in humans as well, females' ovulation is concealed from the males, and often from the females themselves, which gives the females more freedom for their Machiavellian maneuvering.

If rhesus females play their cards right, they will mate with every male in the group and let the alpha male think he's the only male who's having sex, while every other male in the group thinks he's the female's only lover. This will give all the males some confidence of paternity and—who knows?—because they feel they have been chosen by the females to be their secret lovers, they might also think that they have the females' political support.[26] This support could, one day, help them become the new alpha male. Is this Machiavellian enough? There is more.

I once studied a group of macaques with five adult males and thirty or forty females.[27] The alpha male was a big guy named Mr. T, and he was doing his job well. He was a type II alpha male, and his mission in life was to have sex with as many females as possible, as often as possible, without too much concern for what the other guys were up to. Months later, when lots of babies were born in the group, some researchers took the time to do some genetic analyses to see who had really sired all those babies. Well, it turned out that Mr. T wasn't the father of any of them, despite all the sex he had had with everybody. He was sterile and shooting blanks. Three of the other four males had sired most of the infants instead. Even the guy at the bottom of the hierarchy, who lived his life hiding in the shadows and whom I never saw get closer than 5 meters to any female, managed to sire two offspring.

So here is another reason why females don't want to put all of their eggs in one basket. What if there is something wrong with the basket? What would have happened if all the females had been faithful to Mr. T? Once again, rhesus females don't know anything about male sterility, but they are preprogrammed to take this possibility into account. When estrous females have sex with male strangers from other groups, they aren't just shopping for good genes. They are also making sure that the alpha male's sperm isn't the only sperm in the basket with their eggs, just in case. In the jargon of evolutionary biologists, this is called "fertility insurance" behavior.[28] These days, of course, women who have trouble conceiving a child go to the fertility clinic with their husbands instead of the bar around the corner to pick up a cool-looking stranger. But fertility clinics are not an option for rhesus females and have not been an option for women for most of human evolutionary history. Therefore, female brains have been preprogrammed to come up with alternative solutions to the problem of male infertility.

In conclusion, given the way female sexuality works, sex can be big business for both people and rhesus macaques. Because of their biology, rhesus and human females are best positioned to profit from it. The sex business, however, has a complex relationship with political power, and this is the source of an important difference between rhesus macaques and people. In Machiavellian species like humans and rhesus macaques, power is achieved and maintained through political alliances with relatives and friends. Rhesus females play this game much better than males do, and they are the ones to hold the power. Human males seem to have inherited the ability to play the political game well from their chimpanzee-like ancestors, so they play better than females, and they are the ones to control the power. Men have used their power to take over the sex business from women, exploit women, and gain huge profits. Men have learned how to do business with sex on a large scale, with corporations running multimillion-dollar enterprises. So in the end, in our societies, it's men who are profiting big time from the sex business.

Although biology has put a very valuable commodity in the hands of women, they haven't made good use of it. Some women don't even seem to understand that they control a valuable commodity, while others undervalue it. Even the women who run a successful Machiavellian sex business tend to do it on a very small scale. By manipulating one or a few powerful men and getting them to share some of their power, these entrepreneurs seemingly obtain huge benefits for themselves. Unfortunately, they don't share these benefits with other women, and in fact, make things worse for them. More importantly, these benefits are trivial compared with the benefits women could obtain if they used the commodity they control not in order to get men to share some of their power, but to directly obtain power and share it with other women.

Maybe one day women will decide that they want more than good-looking children and a secure and comfortable life, and will set their sights instead on political power. Maybe one day the women who have learned to run a successful sex business will stop playing their own private game and will join with other women to form political parties and big corporations the way men do. On that day, life in human societies will change for good, and we too will become a female-bonded and female-dominated species like rhesus macaques.

PLATE 1 Annual banquet for long-tailed macaques in Lopburi, Thailand. (Photo: Associated Press.)

PLATE 2 Adult female with blonde infant. Some individuals in the Cayo Santiago population carry a gene for blonde hair. (Photo: Dario Maestripieri.)

PLATE 3 A spoof on the film *Pirates of the Caribbean*. (Source: www.worth1000.com.)

PLATE 4 Mother and daughter, both with babies. (Photo: Stephen Ross.)

PLATE 5 On their own: Older juveniles watching from above. (Photo: Dario Maestripieri.)

PLATE 6 "Family tree": Rhesus macaques admiring the landscape on Cayo Santiago. (Photo: Dario Maestripieri.)

PLATE 7 Monkey and bird and their reflections in a pond on Cayo Santiago. (Photo: Dario Maestripieri.)

Chapter 7
PARENTAL INVESTMENT

Investing in Our Future

Mother love is real. And so is father love, for that matter. Parents' love for their children has no parallel in any other kind of love—narcissistic love for oneself, romantic love for other adults, or love for one's own parents. Parents are willing to do for their children far more than they would do for themselves or for any other person in the world. People don't realize this until they have children of their own and find themselves giving up their sleep, time, money, social life, leisure activities, and almost every other thing they used to enjoy. Rhesus macaques don't show many positive emotions such as love or happiness, but if they were to have at least one of them, I would bet all my money on mother love. The social bond between mother and offspring is the strongest you can find in rhesus macaques, and I'd be surprised if such a bond didn't have a strong emotional basis to sustain it. Mother love is strong because it's the product of millions of years of evolution by natural selection. Animal life

FIGURE 17 Mother with twins. (Photo: Dario Maestripieri.)

is all about reproducing itself, and mother love, along with sexual desire, is the fuel that keeps the reproductive engine running.

But there is trouble in paradise. Things can go seriously wrong with mother love. In the book *Mother Love—Mother Hate,* Rozsika Parker reports interviews with mothers who've had the impulse to throw their babies out the window.[1] These are horrible thoughts, and the people who have them feel very ashamed. What's even worse is that some mothers and fathers end up acting on these impulses and do horrible things to their children. You read about them in the news every day. I don't know if rhesus mothers have bad thoughts about their babies because no one knows what kinds of thoughts they have. However, just like humans, some rhesus macaque mothers do some pretty mean things to their babies.[2] The pathology of mother love is not what this chapter is about, however, and you won't hear any more about it here. Instead, what you will hear—which to some of you may sound just as bad—is that mother love can be explained by economics, just like anything else—maybe even more than anything else.

FIGURE 18 Waiting for the mating season. Adult male sleeping while females are busy with babies. (Photo: Dario Maestripieri.)

Our understanding of parenting was revolutionized by an evolutionary biologist named Robert Trivers. His "parental investment theory"[3] was based on the intuition that investing in children is basically no different from investing in the stock market. There are a few differences, of course. When parents invest in their children, they generally hope to gain not money, but lots of grandchildren that will carry their genes through future generations. The cost of parental investment—at least in animals—is measured in the parents' own chances of surviving and producing more offspring. Basically, Trivers's idea was that whatever parents give their offspring—time, energy, or food—leaves them with less of it for themselves and for other offspring. When parents take risks to protect their offspring from danger, they put their own lives and future reproduction on the line.

Parenting is an investment in the real economic sense of the word, and like any other investment, it is regulated by the balance of its costs and benefits. Parents should invest in their children as long as the benefits are greater than the costs. When their balance goes into the red, parents should get out of the business. In animals, decisions about being in or out of the parenting business are not made rationally or with a calculator in hand. Many of them have been sculpted in animal brains by millions of years of evolution. Natural selection has thrown out of the stock market all the players who were making bad investment decisions, so most of today's investors make decisions that are economically sound. People have the same biological predispositions to make economically sound parental investment decisions that animals do, but when the time comes to make those decisions, a lot of other things come into the picture besides evolutionary costs and benefits. So human parents continue investing in their children even when they realize it's a lost cause.

Explaining mother love with economic investment principles sounds very cynical. When seen through the lens of costs and benefits, however, even parents' behavior, not to mention the behavior of their children, will look Machiavellian. Well, Machiavelli's book *The Prince* was about military power, politics, and war, but if he were alive today and wrote a new edition of his book, I'll bet he would include a chapter about power struggles between parents and children as well. He would do well to watch rhesus macaques for a while before writing that chapter.

Getting In and Out of the Stock Market

Some rhesus females have their first menstrual period at two and a half years of age, mate, get pregnant, and have their first baby six months later. Most of them, however, don't get pregnant right away and end up having their first baby a year later. At three or four years of age, a rhesus female is still growing, and she will continue to do so for another year or two. But by this age, rhesus females are very interested in babies and know how to handle them. When rhesus females are about a year old and a new batch of babies are born in their group, they immediately show a lot of interest in the babies. They'll go and check them out—sniff them, touch them, and try to hold them for a few seconds if the baby's mother will allow it. The one-year-old males couldn't care less about the babies. They are just too busy wrestling and chasing other male yearlings. One year later, at two years of age, the difference in interest in infants between females and males is even stronger, and they will differ in this way for the rest of their lives.

Human girls also have an earlier and much greater interest in babies than boys.[4] Some anthropologists and psychologists believe this is because young girls are encouraged by their parents to play with dolls and take care of children whereas boys are given toy trucks and encouraged to become truck drivers, or warriors, or some other male-typical profession. That certainly happens, but there's more to it than socialization. Girls who have a clinical condition called "congenital adrenal hyperplasia," which results from being exposed to extra male hormones before they are born, look more like boys and don't like to play with dolls as much as other girls do.[5] Rhesus mothers don't encourage their daughters to play with babies or discourage their sons from doing it, but the sex difference is still there. And if you give rhesus juveniles a choice between playing with dolls or toy trucks, guess what? The females will choose the dolls and the males will love the trucks.[6] Go figure. If you ever see an adult rhesus male

FIGURE 19 Learning parenting skills: A young female watches her mother and new sibling. (Photo: Dario Maestripieri.)

behind the wheel of a big truck, please stay out of his way and call the police.

The reason young females may be biologically predisposed to be attracted to babies—or to objects that look like them, such as dolls—is that they need to get some hands-on experience before they become mothers. Being interested in babies is probably hardwired in their brains, but being a good mother isn't and requires some learning. So, between one and three years of age, rhesus females watch what mothers do with their babies and try to touch those babies and hold them as much as possible. They don't do real babysitting because rhesus mothers simply don't trust babysitters. Whatever you want to call it, it seems to work, because the females who get more experience with babies make better mothers when their first infants are born.[7]

Having a baby attached to one's chest and carrying it around twenty-four hours a day, however, is very different from playing with one for a few seconds. Some adolescent mothers dump their babies on the ground right after they give birth and just won't have anything to do with them anymore. These babies die unless they get picked up and adopted by another female in the group who happens to have breast milk. Abandoning babies just like that, on day one, is something that only young, first-time moms do.[8] Older females, who have had babies before, don't do it unless they are sick or there is something very wrong with the baby. Not all young females abandon their babies, of course, and those who do, do it only the first time they give birth. When they give birth again the following year, they will keep that baby and be perfect mothers.

Infant abandonment is not necessarily a case of mother love gone bad. Instead, the young mothers who abandon their babies are probably doing the right thing for themselves. Producing milk to feed a baby is energetically very expensive, which means it burns a lot of calories. Some young females who get pregnant right away are still growing themselves and must use all the calories in the food they eat to make sure they become as tall and strong as all the other grown-ups. If they started turning their food into milk for a baby prematurely, their growth could be stunted for good, and that could compromise their ability to have many healthy babies in the future.

Young females who have already grown a lot, or have accumulated a lot of body fat, are likely to have babies early and to keep them. I knew a family of rhesus macaques at the Yerkes Center in which all the females—for at least three generations—happened to be especially fat. All of them had their first babies very early in life and never abandoned one. At three years of age, the females in this family were huge and already looked like adults. I'm not sure if they had a gene for obesity or if they picked up their eating habits from watching their family members. In any case, we know that body fat facilitates early reproduction for people as well. Adolescent girls who are very thin because they are professional athletes, or because they are anorexic and starve themselves, have their first periods later than other girls.[9]

Abandoning a human infant is ethically unacceptable, and when teenage mothers do it they end up in jail. In the rhesus world, however, morality is not an issue. A monkey might be bothered by hearing the cries of an abandoned baby and do something about it, but there is no moral outrage toward the mother, and no one goes to monkey jail. As for the mother, she doesn't seem to have any regrets either. When offspring have low chances of survival or become too energetically expensive, a switch in the animal brain turns off mother love. First-born rhesus infants have low chances of survival for a variety of reasons, and they're potentially very expensive to their mothers because they might interfere with their growth and future reproduction. For some young mothers, the costs are higher than the benefits, so the babies get dumped. Why don't these adolescent females practice sexual abstinence, then, and wait to have their first babies until they are ready? Well, the problem is that the area of the brain that controls their sexual behavior is always pushing it, trying to get the body to start having babies as early as possible. If the body is not ready for motherhood yet—for example, because it's too thin—the body tries to delay or suppress fertility as much as possible. If that doesn't work, it will turn off the mother love switch when the baby is born. To complicate things, the body itself is not completely sure whether or not it is ready for motherhood until the baby is actually born. For example, a rhesus adolescent could be very thin when she has sex in the mating season but be fat enough six months later to be able to sustain lactation. In the uncertainty of the situation, the adolescent body takes a shot at motherhood, and if things don't look promising when the baby is born, it simply says, "Sorry, I guess it was too early after all."

In mice and some other animals, the same switch that turns off mother love also turns offspring into food. So, when things take a bad turn, mothers not only abandon their offspring, but cannibalize them too.[10] I learned this many years ago, when my job as a research technician involved making sure that the mice we kept in the lab had enough food and water in their cages. I once accidentally spilled a bottle of water in a cage with a mother mouse and her pups. As the water flooded her nest—and much to my dismay—the mother started killing and eating her pups. In her mouse mind, the flood was something that made the future look bleak for her pups. By spilling the water and flooding her nest, I altered the balance between the benefits and the costs of parental care: I knocked down the benefits, they became lower than the costs, and the mother took action immediately and got out of the parenting business. She acted like the Americans who pulled out of the stock market the day after Hurricane Katrina hit New Orleans because they knew Katrina would hurt the economy and make their stocks lose their value.

Being the survival and breeding machine that she is, the female mouse didn't waste any time crying over spilled milk, or water in this case, but simply moved her doomed offspring from her love nest to her dinner table. She was already working on her next batch of pups, and a good meal rich in protein and fat could certainly help her survive and make more of them. Female mice typically get pregnant again the day after they give birth, give birth every three weeks, and produce litters of eight to twelve pups. That's why mice are pests and so difficult to eradicate. The mother love switch in the mouse brain gets turned on and off very easily. At the slightest hint that things might be taking a bad turn, mothers are ready to cut their losses and start over again. Rhesus macaques and people take a little longer than three weeks to produce offspring and don't produce twelve of them at a time, so they don't cannibalize their young every time it starts to rain and they can't find their umbrellas. However, that same mother love switch is there, in the rhesus brain and in the human brain.

The Love and Excitement of the First Few Days

It's a beautiful day in the rhesus neighborhood, the sun is shining, the economy is going strong, and a pregnant female is feeling healthy and optimistic and ready to have her baby. The baby finally pops out, she is healthy and hungry, and she looks a lot like her mother and also a tiny bit like the alpha male. Or maybe like the male who's just below him in the hierarchy. Or maybe like that guy who showed up one day and was so charming and persuasive, but then disappeared for good. Never mind.

When things are good the day a rhesus female gives birth, the baby is all she has on her mind. That's all she wants, and she's willing to do anything to keep it. If you take her baby away and give her a baby that belongs to another female, she will adopt it and raise it as if it were her own. She knows it's not her baby, but she'll take it anyway. She wants a baby so badly that someone else's baby is better than no baby at all.

We once did a research project in which female babies had to be swapped at birth between different mothers. Baby swapping, or infant cross-fostering, works only if it's done as soon as possible after a female gives birth, and if her infant is swapped with another infant, preferably of the same sex, born on the same day.[11] One major difficulty with baby swapping is being there as quickly as possible after a baby is born. Rhesus females don't like having babies with people around or when things don't look right. If a pregnant female goes into labor but somebody is standing there watching her, or anything in her surroundings looks abnormal, she will end her contractions, go back to her normal business, and try again a few hours later, or the following day.

At the University of Cambridge and at the Yerkes Center, where I worked, rhesus females typically had their babies at night, on weekends, or on national holidays—any time people were not around. Rhesus macaques can definitely tell that weekends are different from other days of the week because people don't come to work. They also know that these weekends happen regularly. I'm not sure they actually count the days of the week, but they can probably tell that the weekend is coming from the way people around them act (frantic? happy?) on Friday afternoons. These are the kinds of things smart animals learn to figure out when their lives depend entirely on the people around them. Our domestic cats and dogs, and horses like Clever Hans,[12] watch every move we make 24/7 to learn to predict our behavior and figure out ways to get us to give them more food, or anything else they might want from us. When rhesus females give birth, what they want is nobody around, period. Luckily for us, during our baby swapping project at the Yerkes Center, there was at least one person keeping an eye out for new babies every day of the week, so I got a call and had to go to work every weekend and national holiday of that rhesus birth season. With a population of over fifteen hundred rhesus macaques available, we were always lucky enough to have at least two babies born the same day, with the required characteristics, that we could use for swapping.

Back then I was convinced—from talking with other people and reading scientific articles—that baby swapping had to be done within twenty-four or forty-eight hours of birth because that is how long rhesus mothers take to recognize what their infants look or smell or sound like. So I believed that rhesus mothers adopt other females' babies only if they are fooled into thinking they are their own. Then I saw this. A new mother's baby was taken away from her, and another female's baby was put on the floor of her cage. The mother walked up to the baby, looked at it and sniffed it for a second without even touching it, then just walked away and went to sit down at the opposite end of the cage. For ten minutes she acted indifferent and didn't even look at the baby. Then, all of a sudden, she stood up, walked toward the baby, picked it up, and pressed it against her chest. That was it. She had decided to adopt the baby, and from that moment on—for the rest of her life—that baby was hers. That's when I realized that even though we had been trying to swap babies as soon as possible after birth, we were not really fooling anybody. Rhesus mothers, just like human mothers, probably learn to recognize their infants within a few hours—maybe even minutes—after birth. We had confirmation of this when one time, for lack of better alternatives, we swapped babies between two females that lived in the same group. The adoption initially worked out fine for both mothers, but a few minutes after they went back to their group, each mother had her own baby again. They had swapped the babies back between them.

The key to successful infant adoption in rhesus macaques, just as with people, is not recognition, but motivation.[13] People adopt children born to other couples not because they think these children are their own, but because they are highly motivated to have a child. In rhesus macaques, a female's motivation to have a baby, any baby, is very strong after giving birth. It's so strong that sometimes having one baby is not enough. One time, at the Yerkes Center, an old rhesus female and her daughter gave birth on the same day. The following day, the older mother had two babies on her chest. It wasn't a case of twins, though, which are rare in rhesus macaques. The older mother had picked up and adopted a baby that a young new mother in the group had just abandoned. Well, the following day, her daughter had two babies on her chest too. She saw that her mother had two babies and she wanted another one too. We never figured out where she found the second baby. Raising two infants at the same time is very tough for a rhesus female, even if she feels she is Super Mom. Eventually, the older mother managed to do it, but her daughter lost one infant.

These spontaneous cases of infant adoption typically happen within a few days after a female has given birth. After that, maternal motivation begins to fade, and so does a female's willingness to adopt a baby. A week or two after giving birth, the chances of successful adoption are almost zero. But mother love is so strong at the beginning that rhesus mothers, biologically speaking, "make mistakes" and are willing to raise babies that are not their own. The reason for this strong mother love is physiological as well as economic. Maternal motivation is energized by hormonal changes that take place in the mother's body around the time of parturition. In addition, the costs of keeping a newborn baby alive at the beginning are low, while the benefits are high because the mother's parenting behavior can make a huge difference for the baby's survival. As days go by, however, the baby becomes more and more expensive to maintain, and the mother's behavior makes less and less of a difference for its survival. As mother love begins to fade, mothers turn into good accountants and start counting every penny they spend on their babies.

The Business of Parenting

In the first week of their babies' lives, rhesus mothers hold their arms around their babies a lot, let the babies nurse every time they want to, and respond to them every time they cry. If babies are active and try to walk around on their own, the mothers will hold them by the tail and keep them close, or keep a watchful eye on them and retrieve them immediately at the slightest sign of danger. A week or two later, mothers still carry their infants around all the time, but don't cradle them with their arms much anymore, ignore some of their cries, and start rebuffing their attempts to nurse. If the babies wander off, the mothers don't do anything about it, and the babies have to learn to come back on their own. What's happening is that the mothers are already thinking about their next infants and reducing their investment in the current ones.

To be able to have another infant, a rhesus mother has to wean her current offspring quickly. Some people call weaning the moment when breastfeeding ends and an infant starts eating solid food. In reality, rhesus infants and human children don't just stop drinking mother's milk and start eating hamburgers and fries all in one day. Weaning is a gradual process that begins almost at birth. Mothers will walk their children by McDonald's every day, and every day they'll spend a little more time going over the menu with them, explaining what everything is. By the time the children realize that they're not getting mother's milk anymore, they've already been eating Big Macs for a while.

Giving money to a fast food chain is not what helps the mother have her next offspring. The problem is that if the infant's mouth sucks the mother's nipple more than a certain number of times a day, things happen inside the mother's body involving neurons and hormones that prevent her from ovulating and therefore from being fertile.[14] Everybody seems to know that women who have just had a baby are not fertile, but I've heard of women who were shocked because they had sex and became pregnant just a few weeks after giving birth. Having a baby in itself is not a contraceptive. New mothers are infertile only if they breastfeed their babies and do it enough times a day to inhibit ovulation. In scientific jargon, this kind of infertility is called lactational amenorrhea. So rhesus and human females are different from mouse females, and even some other primate females, who can get pregnant the day after they have a baby. Lactational amenorrhea is essentially a Machiavellian trick invented by infants to keep their mothers from giving them a sibling too soon, for reasons that will become clear later.

Rhesus mothers have about six months to reduce the suckling frequency of their infants enough to ovulate and conceive again during the mating season. They must start working on this as soon as possible, as mothers who are too laid back and let their infants suckle any time they want, even in the first month of life, are less likely to conceive another infant during the following mating season.[15] Breeding seasonality puts particular pressure on rhesus mothers because if they miss the mating season, they have to wait another year. So, starting when their infants are two or three weeks old, mothers reject some of their attempts to suckle, and do it more and more frequently in the following weeks and months. When infants come back to their mothers after a little stroll on their own to have a drink of milk, mothers begin to say no and won't let the infants near them. They hold them at a distance with their arms, or turn around and walk away from them, or simply hold their arms across their chest so that the infants can't get to the nipples. If the infants are too persistent, they get smacked. Mothers still spend a lot of time with their infants and help them if they get into serious trouble—they will do that for the rest of their lives—but don't respond to their hunger cries as much and try to cut down as much as possible on their suckling time. Again, the mothers' behavior makes good sense economically. Regardless of their demands, the infants become increasingly capable of surviving without their mothers' help (so the benefits of parental care decline), while becoming potentially so expensive to maintain that not only do they drain their mothers' energy reserves, but also compromise their future reproduction. The same thing happens between human mothers and their babies, but everything happens a lot more quickly for rhesus macaques than for people.

Clearly, the infants are not happy that their mothers are investing less and less in them. The less they get, the more they demand. From the day rhesus mothers start weaning their infants, which is essentially the day after they are born, mothers and infants engage in daily squabbles characterized by indifference, rejection, or outright aggression from the mothers and crying, screaming, and temper tantrums from their infants. These squabbles increase in intensity until they culminate in a real war when the mating season begins and mothers show more interest in having sex than in spending time with their babies.[16] The reason for these squabbles and wars is that mothers and infants don't calculate costs and benefits the same way, and what's best for one is not necessarily what's best for the other. The guy who figured it all out and explained it to the rest of us was, once again, Robert Trivers.

Squabbles between Mother and Offspring

Trivers explained that the squabbles between parents and children that we're all familiar with have a genetic cause—something to do with a thing called the coefficient of relatedness.[17] The coefficient of relatedness is the probability that two individuals share the same genes by virtue of genetic inheritance. It varies from 0 to 1. The coefficient of relatedness between a mother and her child is 0.5 because the child has a 50 percent chance of inheriting any single gene from the mother and a 50 percent chance of getting it from the father. The coefficient of relatedness between two siblings is also 0.5 because siblings have a 50 percent chance of having the same gene (if they both inherit it from their mother or their father) and a 50 percent chance of having a different gene (if one child inherits it from the mother and the other from the father). The probability that children share a particular gene with their grandfather or grandmother, or their uncle or aunt, is 0.25, whereas children are related to their great-grandparents and to their first cousins by 0.125. The following story will help illustrate how the coefficient of relatedness explains the squabbles between parents and offspring.

My friend Jessica just had a baby girl named Sara and loves her very much. Unfortunately, things did not work out between Jessica and Sara's father, and they split up several months before the baby was born. Luckily, Jessica later met a wonderful man, Jeff. They hit it off very well, and plans are already in the works for marriage. Jessica has decided she's going to breastfeed Sara for one year and then stop, so that she can have another child with Jeff. Doctors say that breastfeeding is good for a baby, and that Jessica could breastfeed Sara longer, but she doesn't plan to do so. However, to be fair to Sara, Jessica will breastfeed her next child for one year as well. Jessica will have the same coefficient of relatedness with each of her children—it's always 0.5—so she's going to love them just the same and wants to be fair with anything she gives them, beginning with her breast milk.

So, after eight or nine months of breastfeeding, Jessica begins to give Sara less and less of her milk because she wants the transition to formula to be gradual and smooth. Sara, however, doesn't like this at all. She wants more of her mother's milk, not less. Sure, she will start eating hamburgers and fries one day, but what's the rush? She will want to give up milk when she's good and ready, not when Mother says so. Sara would be happy to have a little sister or brother, but this new sibling will have only one-quarter of Sara's genes. The new sibling will also be Jeff's child and will have half of his genes. Why waste mother's milk on somebody who is 75 percent a stranger when Sara could get all that milk for herself? And this isn't just about milk. Sara is going to have to split everything Mom has—her attention, time, and money—with this half-Jeff child. Mom can have this child with Jeff, and many others too, but not so soon. It would be in Sara's interest for Jessica and Jeff to wait another year at least, if not two or three.

How do rhesus infants and human children try to obtain more than what their parents are willing to give them? This conflict may look unbalanced at first, given that parents are bigger, control the resources children want, and can use physical force to get their way. What are the prospects of victory for the child? Well, Robert Trivers suggested that infants and children use all the Machiavellian psychological tricks at their disposal to get what they want. Rhesus infants and human children pull their tricks from the same bag. Wearing your parents out with constant hysterical crying is so basic it's not even considered a trick. Pretending to be younger and needier than you really are is a good trick. So, instead of showing more independence and more mature behaviors as they grow older, infants and children regress to the behaviors of an earlier age. They become clingier with their parents, talk like little babies again, call for help all the time, and show all kinds of other infantile behaviors. Another good trick is feigning sickness or impending death. After they've been denied their mother's milk, rhesus infants throw themselves on the ground and lie there belly-up, shaking and screaming as if they are having an epileptic seizure, then all of sudden become motionless (but peek at their mothers every now and then to see if they are looking). Human babies have similar strategies. I suspect that what pediatricians call colic—the way some babies cry their heads off for hours every evening in the first two or three months of life—is a trick invented (not consciously but genetically) by some human babies to pretend they're sick and get their parents to do more for them.[18] Not all babies have colic, though, because colic is a risky strategy, and the babies who use it are really pushing it. I recently read in the news that a worn-out mother accidentally killed her colicky baby by putting a lot of vodka into her milk bottle to keep her quiet. Getting into trouble on purpose so as to blackmail your parents into rescuing you is another good trick. Bird biologists believe that chicks beg for food from their parents as loudly as possible so as to attract the attention of predators. The parents have to hurry and give the chicks all the food they can to shut them up before a fox comes and eats them.[19] If bird chicks are Machiavellian enough to blackmail their parents this way, can we expect anything less from rhesus infants and human children?

Rhesus infants also use tactics that aren't so subtle. When a mother tries to have sex and get pregnant during the mating season, her infant may jump on the male's back and try to get him off the mother. Or the infant may squeeze itself between the mother's and the male's bodies and try to push them apart. Infants get very upset when they are momentarily abandoned by their mothers, and some of them, particularly the females, act as if they are going through major depression.[20] Obviously, these infants aren't getting upset or fighting with their mothers over a few extra drops of milk on that particular day. What's at stake is one full year's worth of milk and maternal attention. If the mother gets pregnant, then six months later the infant won't be carried around anymore, and the mother's nipples will be constantly occupied by the new baby. If infants are successful in keeping their mothers from having sex, all the benefits of being an only child will be prolonged one more year.

Trying to keep your mother from having sex is kind of crude, will make her angry, and you might get hurt by the male. So this strategy may be the infant's last resort. It's much better if the infant can sneak a contraceptive inside the mother's body. That way she can have all the sex she wants and everybody is happy. If infants suckle hard in the first six months of life and continue to do so during the mating season, they will suppress their mothers' fertility, and the mothers won't get pregnant no matter how much sex they have.

The earlier infants start working on preventing their mothers from getting pregnant again, the better. In fact, they start so early that they're not even born yet. The squabbling is genetic before it turns behavioral. Not only are infants genetically preprogrammed to squabble with their mothers, they are also preprogrammed to squabble more or less depending on the kind of mother they have. Rhesus mothers who have a biological predisposition to be stingy with their infants produce infants that are predisposed to be very demanding. We discovered this in our study of rhesus infants that were swapped between different mothers at birth. When we studied these infants at six to nine months of age, as they went through their first mating season, we discovered that the intensity with which they were demanding investment from their foster mothers was matched not to the intensity of their foster mothers' resistance (their rejection behavior), but to the resistance of their biological mothers, whom they hadn't seen since birth.[21] Some infants were being very demanding not because their foster mothers were being stingy with them, but because their biological mothers in the group next door were being very stingy. In evolutionary biology this phenomenon is called an arms race. If cheetahs can evolve genes that make them run fast when they hunt gazelles, gazelles are going to come up with genes that make them run faster than the cheetahs. Gazelles without those genes just get eaten. Similarly, if mothers develop genes that make them better able to resist their infants' demands, infants will come up with genes that make them more demanding.

The Thickening of the Infants' Skin

Breastfeeding and carrying infants around are two of the major investments rhesus mothers make in their infants. A third is protection. Protecting an infant doesn't consume as much energy as producing milk or infant carrying but is a lot riskier. Mothers put their own safety and survival at risk when they try to protect their infants. But protect them from what? Aside from predators, the main source of danger to a rhesus infant is other rhesus macaques. Infants are at risk of infanticide from males who might join the group and don't recall having sex with any of its females. Adult females don't commit infanticide, but infants need to be protected from them as well.

On a bright sunny day in Cambridge, a young rhesus infant was walking all alone on a beam three or four meters above the ground. The mother wasn't far away, but she was grooming the alpha male and looked very focused on her task. An adult female from another matriline slowly walked toward the infant, staring at the baby and nervously glancing toward its mother. When she was close enough to touch the baby, she glanced at the mother one more time, then swiftly smacked the infant and pushed it off the beam. I couldn't believe my eyes. The infant fell to the ground with a big thump, followed by an ear-piercing scream. The mother rushed to pick her baby up and ran off with it. Rhesus babies seem to be made of rubber, and this one didn't even get a bruise from the fall. Intrigued by this incident, I decided to study how mothers' grooming activities, so necessary to maintain their political alliances, affected their ability to monitor their infants and ensure their safety. It turned out that when mothers were busy grooming someone else and their infants were walking around on their own, the number of times the mothers glanced at their babies was reduced by half compared with times when the mothers weren't grooming anyone. At the same time, while the mothers were busy grooming, their infants were hurt by other group members more than twice as often as when the mothers weren't grooming. The infants were hurt most often by other females, young and old, who would do all kinds of mean things to them, such as hitting, pushing, pulling, and dragging them. The infants were rarely seriously injured, but always screamed their heads off, when this happened.

At that time I was collecting data on six small groups of rhesus macaques. I watched them every day, including weekends, and soon I began to notice something unusual. Every weekend I heard constant infant screaming coming from one of the groups. The alpha female in this group was an old and mean female named Annette, who had a passion for infant kidnapping. She would snatch a baby from its mother and refuse to give it back to her. Annette didn't have a baby of her own that year and therefore had no milk to offer. Plus, she was very rough with these babies and seemed to enjoy the heck out of it. She wasn't trying to adopt a baby, that's for sure. After one minute in her hands, the baby was already crying and trying to go back to its mother, but Annette, of course, wouldn't let that happen. She held the baby tightly or restrained it by its tail. Rhesus babies may be made of rubber, but if they don't drink their milk often, they will die of starvation or dehydration. After many hours in Annette's hands—sometimes a whole day or longer—the baby was at serious risk, so the animal caretakers had to intervene to catch Annette, take the baby away from her, and give it back to its mother. Annette was left sitting in a cage all by herself for a couple of hours to make sure she wouldn't kidnap the baby again right away, and also as a punishment for her bad behavior. Needless to say, she didn't like that at all. Well, after this happened a few times, Annette learned that the best time to kidnap babies was on the weekend. The animal caretakers were not around, so she could get away with kidnappings and make the infants scream as much as she pleased.

Since then, I have seen hundreds of other infant kidnappings, involving other rhesus females, in other groups, and in other parts of the world. Some kidnappings last only a few minutes, some several hours, and a few of them end up causing the infant's death. Both infant harassment and infant kidnapping seem to be a lot less frequent on Cayo Santiago than in places where the monkeys live in smaller enclosures. I suspect that's true for two reasons. First, on Cayo Santiago, rhesus infants and their mothers have more opportunities to stay away from potential kidnappers. Second, the Cayo Santiago macaques have ways of killing time other than harassing and kidnapping infants; they can run around the island or take a swim. When rhesus macaques are bored and don't have anything better to do, they use their time to stress one another out, and that includes infants too. Rhesus macaques aren't the only monkeys that kidnap infants. There is a population of baboons in Namibia, Africa, in which, at one point, the main cause of infant mortality was infant kidnapping by females.[22]

The infant kidnappings I saw always puzzled me. Although the infants' mothers were obviously bothered by the kidnappings and looked worried about the safety of their infants, they never tried to forcibly take their infants back from the kidnappers. Instead, they would follow the kidnappers around and patiently wait for them to let go of their infants. Sometimes the infants managed to get away on their own, at which point the mothers grabbed them and took off at the speed of light. In some cases, the reason for the mothers not trying a forcible rescue was quite clear: the kidnapper was higher ranking than the mother and she would have kicked her butt had the mother tried anything. Annette, for example, was the alpha female, and she always made that very clear to everyone else in her group. Kidnapping and harassing other females' infants is one of the queen's prerogatives, if she happens to be interested in this kind of activity. I have observed other alpha females over the years who were just like Annette, or even worse. A queen is a queen and deserves respect, but after hours or days of kidnapping, you would think even the most subordinate and fearful mothers would try something to save their babies. Instead, they did nothing. On some occasions they just let their infants die in the hands of the kidnapper. The mothers were passive even when the kidnapper was a lower-ranking female. It's not that they didn't care. They acted like they did. Yet they didn't do anything about it.

A possible explanation for the mother's passivity is the following:

Infant kidnappings that last more than a few minutes are relatively rare, and those that cause infant death are even rarer. Forcible intervention to rescue the infant is risky because the infant could get seriously hurt if a tug of war ensued between the mother and the kidnapper. Sometimes, when a kidnapper tries to snatch a baby from its mother, the mother is still holding the baby. The kidnapper pulls the baby by an arm while the mother is holding its legs. This tug of war is not healthy for the infants, who might get a few inches taller in the process. Because in most cases kidnappings end on their own, mothers don't risk starting a fight, but just passively wait for things to take their course. In the rare cases in which kidnappings last a long time, mothers probably don't realize that their infants could starve or dehydrate to death. They probably don't have a concept of death, or starvation or dehydration, let alone a concept of death from starvation or dehydration.

Having an infant kidnapped to death probably isn't something that has happened to many rhesus females before, so they haven't learned about the risks involved or the appropriate responses. Fatal kidnappings are also too rare for mothers to become genetically predisposed to show the appropriate responses to them. If a mother has a gene for passive responses to fatal kidnappings and loses an infant because of her inactivity, chances are that infant will be the only one she loses to kidnapping in her entire life, so her gene will still be passed on to all of her other offspring. In contrast to kidnapping, rhesus mothers know everything there is to know about aggression and what to do about it. Aggression toward their infants will become increasingly frequent as the infants grow older, especially if their matriline is low ranking. Aggression is something mothers have seen before and know how to respond to. Mothers almost always intervene to protect their offspring when they are under attack by others, and in doing so they expose themselves to retaliatory aggression and risk of injury.[23]

The mother's failure to intervene is not the only puzzling thing about kidnappings. Why do they happen in the first place? Why do females kidnap-to-death other females' infants? Are the deaths intentional or accidental? These are difficult questions to answer. At a basic level, the kidnappers probably just want to hold an infant and behave selfishly about it. They don't care if the infant screams and wants to leave or if the mother looks worried. They care even less if the infant they kidnap belongs to another matriline, or to a female they don't like or don't care about. In some cases kidnappers are very rough with these infants and make them scream and cry as if they are being tortured, which, in a way, they are. This isn't the behavior of a female who has decided to adopt a baby. In the end, I am inclined to explain infant kidnapping the same way I explain other forms of infant harassment. It all comes down to competition between females and the way things work in rhesus society.

Rhesus females that live in the same group compete with one another for food and all kinds of other things they want and need, but mostly for the power that controls everything. This competition takes the form of the unpredictable aggression with which the dominant individuals intimidate subordinates and the constant harassment with which they stress them out. The explanation for infant harassment and kidnapping may be that infants are subjected to the same kind of intimidation and harassment as their mothers, or as subordinate individuals in general, only their treatment is specifi-

FIGURE 20 Infant handling and grunting between two adult females with babies. (Photo: Dario Maestripieri.)

cally tailored to their age and size. Infants clearly start out their social life as subordinate to everybody else in the group. They are not directly attacked because adults seem to have a strong inhibition against attacking young infants the way they attack other adults. Adult females are also inhibited from killing infants directly the way they are inhibited from killing any other group members. They would do it only if their lifetime dominance status were at stake, such as during a matriline overthrow. In those cases the infants are not spared. When hundreds of rhesus macaques were dumped on Cayo Santiago in the 1930s, many infants were killed, and I suspect that adult males weren't the only ones responsible for the killing. Essentially, harassment and kidnapping may be ways to stress infants and, indirectly, their mothers. Therefore, infant kidnapping may be functionally equivalent to the aggression and harassment with which dominant females constantly stress their subordinates.

Infants are loved by their mothers, especially at the beginning, and are very attractive to other females of any age and dominance rank in their group, and not just as potential victims. The social life of the females in a rhesus group during the birth season revolves around the newborn infants, while the adult males simply sit around and wait for the next mating season to start. All rhesus females are crazy about infants and "talk" to them using special vocalizations called grunts and girneys. These vocalizations may be very similar to the baby talk people use when they try to get a baby's attention or to engage it in a conversation. Some females wag their tails while they are vocalizing to a baby, the way a person would dangle a rattle over a baby's crib to get the baby's attention and make it smile.[24] Yet life in rhesus society is stressful for everybody, including infants. Mothers begin rejecting their infants a week after they are born to prepare for their next offspring, and at the same time, infants are already vulnerable to infanticide by males and harassment and kidnapping by females. It takes a very thick skin to survive and be successful in the rhesus world. Rhesus infants start thickening their skin the moment they are born, and everybody around them seems very happy to help in the process.

Chapter 8
THE BUSINESS OF COMMUNICATION

Life in the Social Web

The movie *Waterworld* is set in a distant future in which all land on Earth has been submerged by rising water levels and people live on boats or huge rafts. The inhabitants of Waterworld have all become good sailors and swimmers, nobody suffers from seasickness anymore, and some folks, like the protagonist played by actor Kevin Costner, have even developed gills that allow them to breathe under water. In contrast, the movie *Star Wars* is set in a different kind of future in which people spend most of their time aboard spaceships or hopping around from one planet to another, where they can breathe easily without wearing oxygen masks. In these two versions of our future, people wear very different clothes, but there are always good folks who fall in love with each other and bad folks who want to become rich and powerful at their expense.

People can adjust to life in many different environments, but those environments always have one thing in common: there are lots of other people there. Whether you read history or science fiction, the conclusion is the same: the main source of people's problems is always other people. Even after we figure out how to live on the ocean floor or in outer space, we will still have to deal with far more complex problems: finding friends and mates and fighting enemies. Rhesus macaques find themselves in a similar situation. Adjusting to life in an Indian temple or on a tropical island in the Caribbean is the easy part. Surviving and reproducing in a group with many other rhesus macaques—that's always the problem.

All animal social life involves some sharing of space and coordination of activity—for example, traveling or feeding together—among the individuals that are part of a group. There are many species of animals, however, from fishes to horses, in which individuals live in groups but, most of the time, just mind their own business. These animals aren't even called social, but gregarious, which simply means that they don't mind being in the company of others. In a Machiavellian species like the rhesus macaque, however, the life of every individual is intertwined with the lives of many others into a thick web of intricate connections. Every move a rhesus macaque makes on the social chessboard has a domino effect on everybody else's life, whether they like it or not. Rhesus macaques can't afford to mind their own business. Being passive can be interpreted by others as an encouragement to exploitation. If rhesus macaques just want to be left alone, they must work hard at it. If their goal in life is not just to survive, but also to be successful, they must find a way to get others to work with them or for them. In the Machiavellian social arena, however, everything has a cost, and one has to engage in business transactions with other individuals to obtain anything from them. The business activity with which you get others to do or not to do things to you, for you, or with you is called communication.

In the business of communication, the behavior of other individuals can be purchased with a currency called signals, and all payments are made up front. You flash your signal the way you would flash your credit card, and if it's recognized and accepted by others, you might get what you want. Scams with signals, however, are even easier than with credit cards, so you may want to think twice before you turn your life and all of your belongings over to someone who is sending charming and enticing signals your way. When signals are honest, they are a valid currency for purchasing behavior because they can provide truthful and useful information to other individuals or make them feel good. In both cases, the receiver of the signal will willingly provide the desired behavior in exchange for the signal, so both parties benefit from the transaction, and communication is basically an act of cooperation. Scam artists, however, send signals that convey false information, so that receivers who buy into them are left with nothing useful in their hands. In the business of communication, there are also some rogue players who send signals that make others feel not good, but bad. The rogue players send these signals to others to twist their arms into providing the desired behavior in exchange for stopping the pain. This kind of communication is clearly not an act of cooperation, but of exploitation.

Communication works according to the same principles among humans, rhesus macaques, and all kinds of other animals as well, with the difference that while animals communicate with one another mostly to obtain behavioral responses—they want others to do or not to do something for them—people have invented a form of communication called language, which is particularly good for exchanging information for other information. Providing information through language can be very cheap, and people often obtain information or offer it to others without paying or charging any fees. For example, you can ask a stranger what time it is and get an answer for free. A lot of conversations between people are very cheap business transactions. They are so cheap that they don't even look like business. When we want to purchase a valuable piece of information or some expensive behavior from someone else, however, our payment with signals must be appropriate, or else the business transaction won't work. Animals engage in the business of communication only when it's strictly necessary—to purchase something they really need for their survival or reproduction. In contrast, people engage in communication transactions with one another all the time, and mostly for trivial purposes. But the difference between animals and humans is one of degree, not of substance.

CHAPTER 8

Using Information to Purchase Behavior: Representational Communication

Minnie and Mickey Mouse have just met for the first time at a party in Toontown and have been chatting nonstop for almost an hour. The conversation about the virtues of Swiss cheese versus those of provolone has made Minnie hungry, but she is sitting in a comfortable spot and doesn't want to get up to get a snack for fear of losing her seat. So she asks Mickey to get her a piece of cheese. Minnie would like Mickey to go to the kitchen, take the cheese out of the fridge, and bring it back to her, but why should he do it? Why should Mickey do something for Minnie? Minnie smiles at Mickey when she asks for the cheese, and that's enough for him to do her the favor. Minnie's smile means she likes Mickey and she's going to be nice to him in the future. Knowing something about Minnie's feelings and future behavior is important to Mickey because that information carries the potential for a future gain, so he decides to make an investment and gets her the cheese. Minnie's signal to Mickey happened to be honest, so his investment will pay off (Minnie and Mickey will eventually get married and live happily ever after). Both parties have gained something from the transaction and are happy about it. If Minnie, however, had been dishonest about her feelings and future behavior toward Mickey, his investment would have been wasted, and he would have been exploited.

For Walt Disney cartoon characters and other Machiavellian animals, it's very important to know who your friends are and who your enemies are and what they are willing to do for you or against you in the future. Machiavellian creatures, however, can change their minds very quickly about their friends and enemies, and their behavior is generally difficult to predict. Being able to read minds would give anyone huge power (mind reading has always been my favorite superpower), but luckily for everybody else, this is not an option. An alternative to reading others' minds is simply to listen to what they tell you about themselves and their future behavior. This is not as reliable a source of information as mind reading, but it's still worth something and can be the basis for a business transaction. So one way in which a communication transaction can begin is with one individual sending a signal that conveys information about himself and his future behavior to another individual and hoping that this information is worth enough to get something in return.

Internal states such as emotions and motivations can be good predictors of an individual's behavior and therefore figure prominently in communication transactions. For example, by communicating that we love another individual, we tell that individual that our behavior will be friendly and hope to obtain similar friendly behavior in return. By communicating fear, we tell another individual that our behavior will be nonaggressive and hope to obtain tolerance in return. People and rhesus macaques can communicate many such emotions through their facial expressions. Facial expressions can also convey information about motivation for aggressive, sexual, or friendly behavior. To purchase behavior, one can also offer useful information about other individuals and their behavior, or about particular objects and their location in the environment. For example, rhesus macaques use vocalizations to alert their group members to the presence of food or predators.

FIGURE 21 The expressiveness of rhesus faces: A mildly upset adult female. (Photo: Dario Maestripieri.)

The type of communication in which information is traded for behavior is called representational communication because the signals are used to represent something—internal states such as emotions or motivations, or particular activities, or other individuals or objects.[1] Representational communication is a form of knowledge-based communication because the information transferred through the signal affects the other individual's behavior via his or her knowledge. In other words, Minnie sends Mickey Mouse a signal that represents her feelings, this information affects Mickey Mouse's knowledge of Minnie and her future behavior, and this new knowledge prompts his helpful behavior. People use representational communication all the time, and human language is the ultimate representational tool.

Using Drugs to Purchase Behavior: Nonrepresentational Communication

Selling useful information is only one way to be successful in the communication business. Signals can also make others feel good or feel bad through direct physiological responses similar to those caused by drugs. Some drugs cause physiological responses in the brain that increase pleasure or reduce pain, while others cause discomfort or pain. Making others feel good or bad may be an effective way to get them to do what we want. Signals can also simply make other individuals more or less active, just as some drugs, such as caffeine and amphetamine, increase arousal and attention while other drugs have soothing and relaxing effects. If someone is about to smack you in the head, using a signal that makes that individual less active might work to neutralize the threat. Because the signals that work like drugs don't contain any information or represent anybody or anything, the type of communication that uses such signals is called nonrepresentational communication.

Suppose there is a sound your ears like to hear, such as a particular kind of music. Hearing your favorite tune relaxes you or excites you; therefore, if someone offers this music to you, you are willing to pay for it. At the very least, if you are sitting comfortably in an armchair and the radio starts playing your favorite music, you are willing to get up and turn up the volume. Human babies in every part of the world like to listen to adults who use a particular kind of speech called baby talk or motherese.[2] Baby talk consists of sounds of a particular frequency, organized in particular sequences or melodies. When infants hear baby talk, they turn their heads and look at the baby-talker, smile, and walk toward her, if they can walk. This kind of response is what the baby-talker hopes to get from the baby, so the business transaction is successful and everybody is happy. Another example of a sound we like to hear is laughter. We like to hear other people laughing and so we are willing to work for it—for example, by telling them jokes.

People will also get up from their armchairs if they hear a sound they don't like, this time to turn down the volume or turn off the radio. There are sounds that cause unpleasant states such as irritation, anxiety, and fear. We are willing to do anything we can to avoid them. For example, parents don't like to hear the hysterical cries of a colicky baby and are willing to do a lot to suppress them, such as rocking the baby in their arms for hours or getting in the car in the middle of the night to take the baby to the hospital's emergency room. Other examples of sounds no one likes to hear are screams of fear or pain. Rhesus macaque females may respond to the screams of an infant in pain by picking the infant up and holding it in their arms, but not necessarily because they understand that the infant is in pain and empathize with it. Instead, they might just want to turn off the radio.

Infants are not the only individuals who use nonrepresentational signals. Infants are limited in their ability to represent themselves and the world to other people, and therefore their communicative expressions offer many good examples of nonrepresentational signals. However, adults use them all the time too. Charismatic people, for example, can get others to do a lot for them not because they have a lot of valuable information to offer on the communication market, but because they communicate in a way that makes other people feel good. In contrast, there are many knowledgeable and wise individuals who are willing to provide valuable information to others, but don't do well in the communication business because there is something in the way they communicate that upsets others and turns them off.

Nonrepresentational communication is not knowledge-based, but emotion- or attention-based.[3] Nonrepresentational signals are effective in eliciting particular behavioral responses not because they result in some new and useful knowledge about something, but because they result in specific physiological responses, which in turn affect our behavior. We are able to enjoy music and the visual arts as much as we do not because we have a biological predisposition for music or the visual arts, but because we have a biological predisposition for nonrepresentational communication. This biological predisposition has an ancient history, and we share it with other primates and many other animals.

The nonrepresentational communicative abilities of rhesus macaques are probably quite similar to those of human beings, but their representational communicative abilities are a matter of some debate. According to some researchers, nonhuman primates are very limited in what they can represent to others through signals; they can mostly, or only, represent their own internal states—emotions and motivations—or future behavior.[4] According to other researchers, however, nonhuman primates also have the ability to represent things other than themselves, and therefore have signals that have "external referents," such as particular types of food, or predators, or aggressors.[5] Rhesus macaques probably cannot use complex forms of representational communication in which there is a conventional and arbitrary association between signals and the objects they represent, such as between words and their referents (as in human language). Chimpanzees, bonobos, and other great apes, however, can be taught to use symbolic and linguistic forms of human communication, and according to some, they also do it spontaneously without any explicit training.[6]

Nonrepresentational communication is more hardwired than knowledge-based communication, is processed in different and more ancient areas of the brain, and probably requires fewer and simpler cognitive skills. The sender of a nonrepresentational signal only needs to figure out what works and what doesn't to obtain a certain behavioral response from another individual—which signs and sounds are effective and which are ineffective. The receiver of the signal only needs to know what to do to turn up or turn down the volume. Although individuals are biologically predisposed to produce and respond to nonrepresentational signals, they must learn through experience how to use these signals and respond to them appropriately. In contrast, much knowledge-based communication implies the ability to understand that other individuals have knowledge or ignorance about something. Knowledge and ignorance are mental states, and understanding that others have mental states is part of a complex cognitive ability called a theory of mind. Children develop a theory of mind at about four or five years of age. Whether any monkeys or apes can think about what others are thinking is still being debated.[7]

Knowing the Password Can Save Your Life

Anybody who has spent any time in the military or watched a lot of movies like *Full Metal Jacket*—which depicts life in a military training camp—knows that soldiers don't just wander around and strike up casual conversations with their superiors. Soldiers typically bond with soldiers of similar rank, are harassed by those who rank just above them, and harass those who rank just below them. Sound familiar? Whether U.S. Marines or rhesus macaques, the individuals at the extremes of a hierarchy rarely, if ever, get together and have conversations because those at the top have no interest in doing so and those at the bottom try to avoid it at all costs. If a soldier ever communicates with a general, it is because the general has accidentally met the soldier. Upon meeting the general, the soldier has to come to attention and render the hand salute to remind everyone who is the soldier and who is the general, in case there are any doubts about it.

Subordinate rhesus macaques monitor the dominants' movements all the time and know everything about their walking activities, their favorite paths, the speed at which they walk, and where and when they like to stop for a second to catch their breath. The subordinates use this information to stay out of the dominants' way as much as possible and minimize the chances of a close encounter. In many cases they succeed, with the result that even in a group that lives in a small enclosure, there are pairs of individuals that never interact and communicate, sometimes for months at a time. However, just as there are traffic accidents between groups on Cayo Santiago, there are occasional traffic accidents between subordinates and dominants that live in the same group. Here is an example of how a typical accident may occur.

Blubber was the lowest ranking of six adult and subadult males living in a large group.[8] Blubber spent his days keeping an eye on the other males and staying out of their way. The alpha male, a big guy named Mr. T, vaguely knew of Blubber's existence and had no interest in making his acquaintance. They both knew that proximity between them would be very awkward and that Mr. T would be offended by it. According to the rules and etiquette of macaque society, the default interaction for such proximity would be aggression—the dominant would immediately attack the subordinate. Blubber had learned that lesson a long time ago—when he was in monkey kindergarten—along with the laws of gravity and other basic facts of life.

On a sunny and breezy day, Mr. T was walking along his favorite path with his head down and a serious look on his face, giving others the impression that he might be making plans for the next ten mating seasons, but in reality simply counting his steps, one after the other. Blubber was carefully monitoring Mr. T's movements, but at some point a female with a swollen bottom caught his eye, and he couldn't help but watch her for half a minute, thus losing track of Mr. T. When Blubber remembered Mr. T again, the alpha male was headed his way at full speed and on a full collision course. Blubber had to think fast. He knew that if he didn't move at all, the default response would simply take its course, whereas attempting to flee would only make things worse. The only way for Blubber to avoid the inevitable was to use the secret password that had been passed down to him through many generations within his family and that he had learned early in life—when he was in monkey first grade. Now that the alpha male was getting closer and closer, it was exam day for Blubber, and if he didn't perform well he would end up paying a high price for his distraction.

Mr. T was unaware of Blubber standing right in the middle of his path until Blubber's feet entered the upper part of his visual field.

As the alpha male slowly raised his head to see the rest of Blubber's body, waves of electricity were already traveling along the neurons of his brain to set in motion the chain of events that would unleash the default aggressive response. Blubber, however, had been preparing for this encounter all his life and constantly rehearsing the secret password in his mind. The exact instant Mr. T made eye contact with him, and therefore with perfect timing, Blubber fired his submissive signal—he opened his mouth wide and flashed his clenched teeth right in Mr. T's face. Blubber's performance of the signal was outstanding and earned a standing ovation from his audience. Mr. T was momentarily disoriented by the glaring of Blubber's teeth. It was as if a bucket full of water had been dumped on his firing neurons a nanosecond before the blast. With an acrobatic feat, the alpha male swerved around the subordinate, managing to avoid any physical contact, and quickly resumed his stroll. Another beating had been avoided.

Since Blubber seemed to know his part well, there was no need for Mr. T to reassert his dominance and get his hands dirty. Mr. T knew his part too. He knew how to accept the submissive password from a subordinate and control his aggressive impulses. Blubber had learned in the past that many different signals can be displayed to dominants, some affiliative and others submissive, depending on the circumstances. On that day, a friendly signal would have been inappropriate. What was needed was a submissive signal, and Blubber pulled the right one out of his hat. The choice came down to two signals: the bared-teeth display and the hindquarter presentation. Both are effective submissive signals, and both are recommended for the situation Blubber found himself in: an unexpected approach by the alpha male. The Handbook of the Perfect Subordinate recommends the bared-teeth display if the alpha male is approaching from the front and the hindquarter presentation for approaches from other directions.[9] The Handbook strongly recommends against turning around to display a submissive signal because any movement could be interpreted as an act of hostility and trigger enemy fire. Blubber had carefully memorized this page of the Handbook and made the correct choice. He passed the exam.

Peaceful coexistence in rhesus society depends in large part on the behavior of subordinates. Their avoidance behavior minimizes the chances of casual conflicts with the dominants, and their submissive behavior helps keep the dominants' anger in check. Submissive behaviors are displayed whenever the probability of receiving aggression rises above the baseline for any reason. Proximity is one such reason; the presence of food and any fighting activity in the group are other precipitating factors. Fighting between any two individuals raises the probability of all low-ranking animals receiving aggression. In these situations, the likely scapegoats start firing submissive signals left and right. Submissive signals are also occasionally used in a preemptive manner, when there is no discernible risk of aggression, just to appease dominants. A subordinate will walk up to a dominant who's minding his own business, show the bared-teeth display or the hindquarter presentation, and then walk away. Rhesus macaques, however, perform this unsolicited appeasement behavior a lot less than other species of macaques[10] and possibly even humans. In rhesus macaques, approaching a dominant, even for appeasement purposes, is risky, and subordinates just don't want to take the chance.

Regardless of all the avoidance, submission, appeasement, and other preventive behaviors used by subordinates, aggression occurs anyway. Sometimes aggression is just what the dominants want to do, and once their minds are set, nothing will keep them from doing it. In the one-minute time window after a subordinate has been attacked by a dominant, the risk of additional aggression is at its peak.[11] The subordinate risks being attacked again by the same aggressor as well as by other individuals who may join the fight. This is the time when subordinates must unleash all of their communicative potential. They face their aggressors and bare their teeth to them while raising their tails as white flags to signal surrender to anybody else in the vicinity who might be considering joining the fight. The subordinates also scream their heads off to solicit support from their family members. Sometimes these signals are effective in preventing further aggression or in recruiting support, and sometimes they aren't, for reasons that aren't always clear.

But why do these submissive signals work at all? To a spectator watching the scene, the encounter between Blubber and Mr. T looked very much like the accidental encounter between a general and a soldier in a military training camp in which the soldier salutes the general. The general would get angry if he didn't get the salute from the soldier, and the soldier would get into a lot of trouble for not saluting. Peaceful coexistence in the military depends on saluting superiors as much as peace in rhesus society depends on submissive signals from the subordinates. In any social system with a strong hierarchy, individuals need to constantly remind each other and everybody else of their relative ranks. Submissive signals are effective in keeping the peace because they do exactly what fighting does: they determine a winner and a loser, but without the energy expenditure and the risk of injury that fighting entails.

A subordinate may use a submissive signal to reassure a dominant that things between them are clear and there is no need for another fight to determine who is dominant and who is subordinate. In this view, the meaning of the bared-teeth display might be something like "I know I am subordinate." It's an acknowledgment of status.[12] This would explain why the signal is effective in inhibiting the aggressive behavior of the dominant. The reason a signal is effective in eliciting or suppressing a particular behavioral response, however, may not be the same as the reason individuals use it. Animals—and people too—use signals effectively without necessarily being aware of their consequences. If everybody had to think ahead about how another individual might or might not respond to a signal, every communicative interaction between two individuals would have to be accompanied by a lot of thinking. And some of this thinking would have to be quite sophisticated, at least for a rhesus macaque. If a subordinate consciously used the bared-teeth display to acknowledge his status to a dominant, that would imply that the subordinate was aware that the dominant had knowledge or ignorance about something and that the signal could affect this knowledge or ignorance.[13] When someone says to me, "I want you to know that . . . ," that person must understand that I don't know something, otherwise he wouldn't say it. Knowledge and ignorance are mental states—they exist only in one's mind—and as we have seen, rhesus macaques don't seem to have the ability to think about what's in the mind of another individual.[14]

A simpler explanation for the occurrence of the bared-teeth display is that this signal is either an expression of fear—"I'm scared"—or a plea—"Don't hurt me!"—or a combination of both. By seeing fear in the subordinate or hearing his plea, the dominant is reassured that his status is not being challenged and that no fighting is needed, so the outcome is the same as if the subordinate were saying "I want you to know I'm subordinate." I like this simple explanation for the bared-teeth display because rhesus macaques also show this signal to people, to snakes, and even when they are frightened by a loud noise.[15] In the military, the hand salute is a real acknowledgment of status. A soldier would salute a general, but he wouldn't salute an attacking tiger. He certainly wouldn't give a hand salute in a movie theater when he sees Freddie Krueger on the big screen, slashing one of his victims with his razor-blade fingers. I suspect, however, that a rhesus macaque watching a Freddie Krueger movie would show a lot of bared-teeth displays to the big screen.

People have a facial expression that looks almost exactly like the rhesus macaque bared-teeth display: the smile. Although human smiles sometimes serve a submissive function—that is, subordinate individuals smile to dominants to inhibit their aggression (often verbal or psychological, but also physical)—social smiles communicate a generally friendly motivation and not just fear or submission. As in the Minnie and Mickey Mouse example, a smile can mean, "I like you and I am going to be nice to you." Social life outside the military is less formal, more relaxed, and generally friendlier than life in the military. Although every social relationship between two human beings probably has a dominance component, dominance doesn't play as large a role in our everyday social lives as it does in the military and in rhesus society, so communication about dominance is less frequent and more subtle. More importantly, life in civilian societies entails much more contact and coexistence with strangers than life on a military base or in a rhesus group. Communication of a friendly attitude can increase tolerance between strangers or alleviate tension between people who must work together but don't know each other well, so during the course of their evolutionary history, humans modified the bared-teeth display they inherited from their primate ancestors into a multipurpose friendly signal, the smile.[16]

Threats, Bluffs, and Poker Games

A close encounter with an alpha male is only one of many circumstances capable of triggering submissive signaling from subordinates. Another powerful trigger is a threat. Rhesus macaques threaten others by staring at them, raising the eyebrows and scalp, opening the mouth wide without showing the teeth, and making all kinds of harsh sounds, including grunts and screams. A threat conveys information about an individual's physical strength as well as his or her aggressive motivation and willingness to engage in a fight, so being threatened clearly qualifies as a situation with a high risk of aggression. In rhesus society, all threats are directed down the hierarchy, from higher-ranking to lower-ranking individuals, just as all submissive signals are directed up the hierarchy. Clearly, individuals who threaten others are not necessarily ready to fight; otherwise, they would just go ahead and do it. Using threats to settle a conflict is like playing poker. The threat conveys information about your physical strength and your willingness to fight, and these are the tokens you put on the table. If the other player is impressed by your display of confidence, that will do it, and you are the winner. If the other player is not impressed, or is very confident in the quality of her own cards, then you'll have to put your cards on the table and fight. Obviously, bluffing is as important in threats as it is in poker, and detecting a bluff is not easy if the players are good.

If a rhesus female at the bottom of the hierarchy is threatened by the alpha female, she will respond with a submissive signal because her cards are bad and her self-confidence is low. In the poker game played by rhesus females, players with many family members have good cards and those without them don't, so there isn't much room for bluffing. The winner and the loser of the game will be the same every day until the demographics of the group change and the subordinate matrilines outnumber the dominants. On that day, a subordinate female will respond to a dominant's threat with another threat, all cards will be on the table, and there will probably be a matriline overthrow. In the poker game played by males, however, winning a fight depends more on physical strength and aggressive motivation than on support from others, and the cards in one's hands are not so

FIGURE 22 Adult female showing an open-mouth threat. (Photo: Dario Maestripieri.)

obvious. Winning some fights at the beginning will buy a male alpha status, and he will be able to maintain it with lots of threats and little fighting for a year or two, while his victories are still fresh in the others' memories and therefore his threats are credible. As time goes by, however, and their memories of the fighting alpha male begin to fade, other males will increasingly ask to see his cards. One day another male will put better cards than the alpha male's on the table, and the game will have a new winner.

Rhesus macaques play dominance poker games not only with one another, but with people as well. They seem to understand that people work well as a team when it comes to fighting, so they generally won't challenge groups of people even if they outnumber them. When a person is alone, however, things are different. When I walk around on my own on Cayo Santiago, every now and then I am confronted by an adult male who wants to play poker with me. Obviously this guy and I haven't fought before, so we don't have an established dominance relationship. He is eager to settle the issue so that if he wins, he can go around and brag to the other monkeys that he has outranked a researcher with a Ph.D. The rhesus male can obviously tell that I am bigger and presumably stronger than he is, so some of our cards are already on the table. Rhesus macaques, however, have learned that some people are afraid of monkeys. They have also attacked people in the past and gotten away with it. Because fear and willingness to fight are not something you can see up front, that gives the rhesus male some room for playing his hand. So he makes his bid and threatens me.

At this point I have two options. I can say to him, "No thanks, I am not interested in playing dominance poker games with monkeys," or I can play and make my own bid. I always choose the first option, so upon receiving the threat, I simply avoid making eye contact with the monkey and keep minding my own business. I must be careful, however, not to show any response that might be interpreted as avoidance, fear, or submission, because that would escalate the game—in the wrong direction. If I give the monkey a hint that he might be able to outrank me, he'll go for it. Most of the time, my indifference works and gets me out of the situation, but every now and then I run into a stubborn male who won't take no for an answer and keeps on threatening me. At that point I have no choice but to play, and play well. When I make my bid, I must make the monkey believe that if he makes me put my cards on the table, he will lose everything he's got. So I stare him down, put on the best angry face I am capable of, and yell at him from the bottom of my lungs. There are enough similarities between the expressions of emotion and motivation of people and rhesus macaques for the two species to understand each other without the need of an interpreter. Obviously, I have no intention of wrestling with a macaque, and I'm not so confident I would win, but my bluff is good enough for him, and he finally leaves me alone.

When I was in middle school, I knew a bully who seemed to know a lot about rhesus macaques and how threats work. He was big and tough and had established his alpha male status among all the gangs in the neighborhood through fighting and various acts of bravado. To increase the number of people he outranked, as well as to test and discourage potential challengers to his alpha status, he would spend his days standing on a street corner and staring people down. If anybody made eye contact with him, he would approach the person and make more explicit threats to escalate the confrontation and to test how far the person was willing to go. The bully was a violent guy who had been in and out of jail a lot, so he wasn't really bluffing, he was the real thing. The day I became the target of his gaze, I did what I do with rhesus macaques. I avoided eye contact with him and said no thanks to the game. My choice was wise, and luckily for me, it was good enough for him too.

Friendly Signals

The exchange of threats and submissive signals accounts for a large proportion of the communicative interactions between rhesus macaques. Dominance occupies such a central role in the social lives of rhesus macaques that there can't be any misunderstandings about it, so communication of dominance and submission is continual and redundant. Rhesus macaques' obsession with power and their constant struggle for dominance is only part of the story, however. The other part of the story is their temperament. Rhesus macaques get excited very easily, and nothing makes them angry or scared as easily as social stimuli such as faces. Look a chimpanzee in the eye and you will get a calm look in return that seems to ask, "What's up, buddy? What can I do for you?" Look a rhesus macaque in the eye and you might get the "You talking to me? You talking to *me*?" response. Look another rhesus macaque in the eye and she might freak out and bare her teeth, scream, and beg for her life. Rhesus macaques give these kinds of responses to people, to other rhesus macaques, and to their own images in the mirror. They are very expressive, and you can always read their emotions in their faces, which is one of the reasons I enjoy watching them. Rhesus macaques can also read emotions in one another's faces, and their excitable temperament is a powerful catalyst of social behavior.

If one puts together the notion that aggression is an integral part of rhesus macaques' social lives with the fact that they are very excitable and chronically stressed, the result is a view of rhesus soci-

FIGURE 23 Expression of fear in a young adult female. (Photo: Dario Maestripieri.)

ety as one in which individuals walk around with a loaded rifle and are ready to fire at the slightest perturbation. Rhesus macaques look comfortable and relaxed only in the presence of their closest kin because they know them well and their behavior is highly predictable. Sitting, walking, and feeding next to one's closest family members is relatively safe and doesn't need to be negotiated with complex signaling. The social bonds with relatives are also reinforced with grooming and huddling, and sometimes related females hug each other the way people do. Things are different, however, between individuals that belong to different families or are very distant in rank, such as the monkeys at the top and the bottom of the hierarchy. These individuals have poor relationships because they interact rarely and those interactions tend to be negative. Therefore, when subordinates find themselves within an arm's reach of the dominants, they always prepare for the worst. They either avoid proximity to dominants altogether or they negotiate it very carefully with signals.

If subordinates want to do business with dominants, such as offering grooming in exchange for tolerance or agonistic support, they approach the dominants very carefully and signal their intentions to them with friendly signals such as lip-smacking or soft and harsh vocalizations called grunts. Lip-smacks and grunts may also continue through the grooming session just to keep the dominant quiet and happy. When dominants want something from subordinates, they too must signal their intentions. If the alpha male wants to have sex with a female from a low-ranking family, he can't just walk up to her and try to mount her, because she would freak out and flee, screaming for help. The male instead introduces himself by lip-smacking to her or by pursing his lips and retracting his scalp, making a face called the "pucker." The male will even go so far as to bare his teeth to the female to reassure her that this is about sex, not violence. Finally, the male may also perform a quick mating dance to get close to the female little by little, without freaking her out. He will take a couple of steps toward the female, then turn around and take a couple of steps back, then turn around again and make his final approach. The male is literally making a pass at the female, and you can think of his behavior as a courtship ritual. When we think about male sexual courtship, we usually think about efforts a male can make to look attractive to a female. In animals in which males are larger than, stronger than, and potentially dangerous to females, however, one important function of male courtship is to reassure the female that the male means no harm and that the business he has in mind is only sex. There are also species, such as some spiders, in which it's the males who fear the females because males are much smaller than females and happen to be one of their favorite foods. In this case, the function of male courtship is to appease the female, to try to get her out of the mood for eating, or at least give the male the time for a quick copulation before he gets swallowed.[17] This courtship ritual works best if the male brings the female a snack to keep her busy eating while he is copulating with her, and males in a lot of insect species do just that.[18]

Rhesus macaques have one main friendly facial expression, lip-smacking, and one main friendly vocalization, the grunt, that they exchange when they get together for grooming or sex. These signals seem to have a generic friendly connotation and can be used in many different contexts and circumstances. Macaques use lip-smacking not only to let someone know they are coming close for grooming or sex, but also to request it from someone else. For example, if the alpha male is sitting and minding his own business, and all of a sudden he makes eye contact with and lip-smacks to a female who's sitting or standing even at a great distance from him, the female will rush to the male and start grooming him. Judging from his contented and relaxed expression, that's exactly what he wanted. But how did she know? Obviously, lip-smacking in itself doesn't mean "I want to be groomed," so the female must have thought something along these lines: "The alpha male is acting friendly to me, but he's just sitting and doing nothing, so I don't think he wants to have sex with me or groom me; I guess he wants me to go over there and groom him." The male may also give the female a hint that this is about grooming by beginning to lie down and exposing the part of his body he wants to be groomed, whereas if he had been asking for sex, he would have touched the female on her hips when she got close to him.

Rhesus females lip-smack and grunt to other females and their infants a lot. It's not clear whether a female who wants to touch an infant grunts to the infant's mother to let her know of her friendly intentions, or whether she grunts to the infant to get the infant to do something, or whether she just grunts because she is excited to see an infant, or all of the above. Rhesus females also make a nasal and melodic sound, called a girney, when they get close to an infant and seem interested in touching and holding it. The girney may be a kind of baby talk that works to get the infant's attention and possibly a response as well.[19] The female may also wag her tail while looking at a baby and making grunts and girneys.

For some reason, mothers never make grunts or girneys or wag their tails to their own infants—they do it only to the infants of other females. Mothers lip-smack or bare their teeth to their infants when the infants are walking around on their own and it's time for them to come back. I have also seen mothers walk backward in front of their infants and lip-smack to them to encourage them to walk and follow.[20] It's one of the most humanlike behaviors I've seen rhesus macaques do, and I was very impressed by it when I first saw it. I have never seen a rhesus mother play with her infant, however, or give her infant a piece of food. Instead, I've often seen a mother take a piece of food out of her infant's mouth and eat it herself. Rhesus mothers can be very protective of their infants, but they don't waste time on random acts of generosity.

Chimpanzee mothers tickle their infants, and the infants respond with facial expressions that look like smiles and laughter. Rhesus macaques don't do any of that. In twenty years of watching rhesus macaques, I have never seen anything that resembled an expression of feelings such as love, compassion, or guilt, or positive emotions such as joy or happiness. The closest they've gotten to an expression of pleasure is their relaxation when they lie down while being groomed by another macaque. When rhesus males have sex and ejaculate, they bare their teeth and squeak. I am sure they experience sexual pleasure the same way we do, just as they experience pain the same way. To be fair to them, rhesus macaques are not unusual among animals in not having clear expressions of positive emotions or feelings. The clearest expressions of emotions in animals, whether those expressions are behavioral or physiological, are those of negative emotions such as fear and anxiety.

Life in the Monkey Army

As impressed as I am with the complexities and subtleties of rhesus macaque social behavior, I can't honestly say that I'm impressed with their communicative abilities. Only a few different vocalizations are used by rhesus macaques, and they are far less complex than the vocalizations of other animals, most notably birdsong. Rhesus macaques have relatively few facial expressions and body postures, and these appear to have a rather generic meaning, if they have a meaning at all (in the sense that words do, for example). In addition, rhesus macaques don't use their hands to make gestures the way people do. People communicate with one another constantly, whereas one can watch a group of rhesus macaques for quite a while and not see a single signal—facial expression or vocalization—being exchanged. When they do talk to one another, the content of their conversations is relatively simple. Rhesus macaques have vocalizations that alert others to the presence of food, predators, or aggression. All these vocalizations have an emotional or motivational connotation, but may also convey more precise information, such as the quality, quantity, and location of the food being found, whether the predator being encountered is an animal with four legs or a bird, and whether someone's aggressor is male or female, kin or nonkin. Other than that, rhesus macaques don't seem to talk about the weather, how green the forest looks today, or any other aspect of the physical environment in which they live.

One of most salient aspects of people's environment is the presence and activity of other people, and this is reflected in their conversations. For example, studies done in college cafeterias have shown that 58.3 percent of the content of students' conversations is gossip. Anthropologists see the same pattern in the villages of nonindustrialized societies.[21] Most rhesus macaque communication is also about the social environment, but they don't gossip about others the way we do. Instead, most of their communication is about social behavior: the things they would like to do, the things they would like others to do or not to do, and the things they and others could or should do together, such as sex or grooming. Even in these kinds of conversations, however, rhesus macaques appear to be quite limited and restricted, not only in comparison with humans, which is obvious, but also in comparison with other species of macaques closely related to them. For example, pigtail and stumptail macaques appear to have a wider repertoire of visual and vocal signals than rhesus macaques, and they use those signals more often and in more complex interactions than rhesus macaques do.[22]

How animals communicate with one another depends on their anatomy and physiology; that is, on their bones, muscles, vocal cords, brains, and the olfactory receptors in their noses. How much animals communicate with one another and what they talk about depends in part on their biology, but also on the environment in which they live. The anatomy and physiology of rhesus macaques is virtually identical to that of other macaque species, and until very recently, the forests and other habitats in which they live have been very similar as well. So if rhesus macaques have a simple and unsophisticated system of communication relative to other macaques, the reason must have something to do with their social environment.

The organization of rhesus macaque society and the way it functions may keep social communication to a minimum. Consider the following. Communication is an adaptation for social interaction, and in rhesus macaques females are generally more sociable than males. Male and female infants initially grow up in a highly social environment, but then, whereas females remain embedded in their large and complex social networks, the males adopt a semi-solitary lifestyle. After emigration, males spend a lot of time on their own or with a few buddies. When they join a new group, they participate in social activities only if these activities involve sex or alliance formation—that is, opportunities to move up the hierarchy. For the most part, they just hang out on their own. When it comes to communication, rhesus males are not very articulate or subtle. Rhesus females, on the other hand, are very sociable, but they spend most of their time with their closest family members. That is, they hang out with individuals they know very well, with very similar social interests, and whose behavior is very consistent and predictable. Females depend on their close relatives for support, and they do a little business on the side, exchanging sex for protection with the resident males. They also have sex behind the bushes with strange males. These are not social circumstances that encourage intense and complex forms of communication. Females' social lives are constrained by the rigidity of the matrilineal system, and so are their opportunities for communication. Interactions between females from different families are constrained by the generally hostile relationships between matrilines and by the rigidity of the hierarchical system, which makes it difficult for individuals distant in rank to interact extensively and affiliate. A rigid hierarchy doesn't encourage complex and sophisticated forms of communication, only clear and redundant signals that communicate differences in rank between individuals. Complex political and business interactions across the boundaries of kinship and power classes would encourage sophisticated communication and negotiation, but rhesus societies can offer little of that. Rhesus societies are organized like armies, and the function of armies is to make war, not to produce language or culture.

Soldiers socialize mostly with other soldiers of similar rank. Among those other soldiers, they especially bond with people who are from the same part of the country, have similar backgrounds, or share similar experiences and interests. In other words, soldiers re-create small kin groups and form kinship-like bonds with those individuals who are most similar to themselves. In some cases, the kinship groups are real. While I was doing my military service in the Italian Air Force, I once opened the phone book of my barracks and found on almost every page long lists of people with the same last name. In Italy, many men from the same family enroll in the military over generations, and through various nepotistic mechanisms, they all end up at the same base. Judging from the phone book I consulted, my Air Force base had a social structure similar to that of a rhesus macaque group. There was a strict and linear dominance hierarchy and large kin groups similar to rhesus matrilines, but consisting of men instead of women. Whether the kin groups are real or virtual, nepotistic behavior in the military is rampant. Soldiers belonging to the same group share camaraderie and mutual support and jointly harass soldiers from other groups. New recruits receive the same kind of treatment as female monkeys that are forcibly introduced into a rhesus group, and if they survive that treatment, they start out their careers at the bottom of the hierarchy and have to work their way up. Thus, the army has something akin to the hierarchy and matrilineal structure of rhesus society, and the dynamics of communication may be similar as well. Communication within the army is not meant to be complex and sophisticated, but clear, simple, and effective, so that when the army is at war, soldiers don't question orders or waste their time negotiating their duties and responsibilities. Armies have a job to do, and they do it well. And so do rhesus macaques.

Chapter 9
MACACHIAVELLIAN ORIGINS OF LOVE AND COMPASSION

Drivers and Passengers on the Evolutionary Journey to Large Brains and Complex Intelligence

Homo sapiens is a political animal, and so is *Macaca mulatta,* the rhesus macaque. In both human and rhesus macaque societies, it's difficult to survive and be successful without the help of others. Success depends on political power, and that power is acquired and maintained through the formation of alliances with other individuals. In rhesus macaques, most political power is held by females through kinship-based networks of alliances: the matrilines. Humans cooperate and form alliances with their family members too, but given the size and complexity of human societies, the pursuit of political power on a large scale requires the formation of networks of alliances with unrelated individuals. These networks of alliances are called political parties, and in virtually all human societies, they are controlled by males.

Wherever there is cooperation there is also competition. Individuals cooperate with one another to compete against others. Because of competition, the main source of problems for people is other people, and for rhesus macaques it's other rhesus macaques. To deal with the complexities of cooperation and competition, rhesus macaques and people have evolved a sophisticated and opportunistic form of social intelligence. This Machiavellian intelligence is also observed in other primates—for example, in baboons and chimpanzees—and in some other animals as well, but not in all of them. It's not that non-Machiavellian animals are stupid, but they have different kinds of social organizations, and their social intelligence is probably adapted to their lifestyle and needs. Mountain gorillas, for example, live in small groups consisting of a large male and his harem of females and children. A successful male in mountain gorilla society is a strong and quiet type, and a successful female is one who can find such a guy and stick with him.[1] These personality traits don't encourage the nurturing of political ambitions. The lifestyle of the mountain gorilla would produce someone like King Kong, but not Machiavelli.[2]

Life in large social groups may encourage not only the evolution of Machiavellian social intelligence, but of complex intelligence in general. Humans have the largest brain relative to their body size among all the primates, and the great apes tend to follow suit.[3] Humans have an especially large neocortex, the part of the brain responsible for our most complex social-cognitive skills. There are many different theories as to why and how people and other apes evolved a large neocortex. According to the Machiavellian intelligence theory, our large neocortex is the result of living in large groups, in which an individual needs to remember a lot of different faces, form alliances, and keep track of what everyone else is up to.[4] Consistent with this idea, primate species that live in larger groups have a larger neocortex than species that live in smaller groups, other things being equal.[5]

There is an interesting twist to this story. According to current primatological theories, whether or not primates live in groups is determined by females and their needs.[6] If females are better off looking for food and avoiding predators on their own, then the species as a whole has a solitary lifestyle. If females need the help of males to raise their offspring, then the species lives in pairs or small family groups. Finally, if females need to cooperate with other females to find and defend their favorite food or to protect themselves from predators, then the species lives in groups. These groups will be large or small, again, depending on the females' needs. Males have needs too, of course, and those needs are called females. There are some very brilliant minds among the scientists who study primate social evolution, and after decades of intellectual effort and countless mathematical models, they have concluded that the contribution of males to the evolution of primate sociality is best summarized in the following statement: males go where the females are.[7] So if the females are solitary, the males just follow them around, and if they live in groups with other females, males join those groups. Males' primary goal in life is always one and the same: sex. Because males eat a lot of food, are potentially dangerous, and don't help much with the kids, females of many primate species tolerate only a few males within their groups and expect them to make themselves useful by fighting predators and primates from other groups.

The interesting twist in the Machiavellian intelligence theory is that the association between neocortex size and group size seen across many primate species is seen in the females and not the males.[8] In other words, the more females that live in the company of other females, the larger the neocortex of the species, whereas male group size does not correlate with neocortex size. This interesting finding suggests that the evolution of complex intelligence in Old World monkeys and apes, including humans, may be due to the increasing complexity of female social life. In the evolutionary journey that led to big primate brains and complex intelligence, females and males traveled together and eventually reached the same destination, but females were the drivers and males were the passengers. Smart females produce smart kids, and some of them happen to be male. Males are genetically and anatomically similar to females, so as females became smarter, males—at least some of them—became smarter too.

Inequality of the Sexes and Male and Female Power

Equality of the sexes is rare in primate societies. Typically one sex has more political power than the other, and the rules that govern the society are made by the sex with more power. Complex primate societies are dominated by the sex that is better at forming alliances, and in most such societies, that sex is female. It is kinship that provides the basis for strong and long-lasting alliances, and, because in most primate species males emigrate and leave their families behind, females have more opportunities to form alliances with relatives than males do. Species in which females emigrate and males form strong bonds with their male relatives are the exception among primates. Chimpanzees are one of those exceptions, and among chimpanzees, males hold the political power. Humans probably evolved from primates similar to chimpanzees, in which females emigrated from their group while males stayed and formed political alliances with other males.[9] As a result, human societies have always been male-dominated. If human societies had retained the predominant primate pattern of female bondedness and female domination, things would have been different, and rhesus macaques give us a glimpse of what they might have been like.

In rhesus macaques, males' emigration from their natal group prevents them from establishing alliances with their fathers, brothers, or any other family members. When males immigrate into a new group and reproduce, they will probably remain there for less time than it takes for their offspring to reach adulthood. If they are successful, or lucky, and remain longer, they will kiss their sons goodbye anyway when it's their sons' turn to emigrate. Adult males have opportunities to form alliances mostly with unrelated individuals and during relatively brief periods of their lives, so these alliances tend to be unstable and short-lived. Thus rhesus males are pretty much on their own in the political arena. To make it to the top of their social world, they need the physical strength and the motivation to fight other males and the opportunistic social skills to exploit other individuals and situations. The life of a rhesus male is a ladder, with alpha status at the top. Every male spends the best part of his life climbing the ladder, having the bottom of the guy above him right in his face and the guy below breathing on his neck.

Alpha male status comes with a lot of benefits, such as unrestrained sexual activity with all the females in the group and priority of access to food and other resources. The power of an alpha male, however, is mostly representative. He doesn't make the rules and control society. A rhesus alpha male has the social and sexual privileges of a king or a president, but one in a government in which the political power is in the hands of the prime minister and her political party. In rhesus society, these are the alpha female and her matriline. The power of the alpha male doesn't have a solid base of support, and standing all alone at the top of the ladder can be a precarious situation. There are always other males climbing and shaking the ladder, and the moment the prime minister decides to switch allegiance to one of them, the king is set for a big fall. So the few males who make it all the way to the top will be alpha males for a few years before they get knocked down by another guy, and from then on their lives will just go downhill. All they can do is make sure their star shines as brightly as possible while it's up in the sky and hope that the offspring they managed to sire during their fifteen minutes of fame will take good care of their genes.

Female power is different. Females are in for the long haul. They base their power on support received from female relatives, and they acquire, maintain, and nurture that support with lifelong social relationships. For females, power is in numbers: large families overpower small families. Female power is built over many generations and can last for many generations more. Rhesus females allow one male to be their representative leader and tolerate a few others around, but they are the ones to control the political scene. Their matrilines compete with one another for political power within the group the way political parties and the big corporations that support them do in modern capitalistic societies, or the way families compete for and control power in the Mafia world.

Nepotism and Despotism in Rhesus Macaque Society

In female-bonded and female-dominated species like rhesus macaques, the laws are made by adult females, meaning that the social dynamics are shaped by the nature of female social relationships.[10] In other words, the way adult females treat one another affects the way they treat other individuals; for example, the way adult females treat infants or the way females interact with adult males. It may also affect the way other individuals treat one another; for example, how males interact with other males. The fact that rhesus females have mostly antagonistic relationships with females from other matrilines probably has something to do with the fact that adult males generally have poor and antagonistic relationships with unrelated adult males. In other female-bonded species, such as baboons, and even in other macaque species, unrelated males do form coalitions with one another against other males or females, and these coalitions may involve real cooperation or reciprocation of favors between males.[11] In rhesus macaques, however, male-male coalitions are relatively rare and are usually accounted for by individual opportunistic behavior without any real exchange or reciprocation of altruism. Rhesus females seem to have their own reasons for treating one another the way they do, but their social dispositions also rub off on the males. As with intelligence, females are behind the wheel of social relationships, and the males simply tag along.

The social relationships of rhesus females are defined by kinship and dominance. Rhesus females treat relatives very differently from nonrelatives, and they treat higher-ranking females very differently from lower-ranking females. There are many other primate species in which females treat kin and nonkin differentially, but not to the degree that rhesus macaques do. Many other primate species also form dominance hierarchies, but differences in dominance between individuals don't affect their social interactions as much as they do in rhesus macaques. Relationships and societies in which the behavior of individuals is very biased toward kin and against nonkin are called nepotistic, whereas those in which the influence of kinship is minimal are called individualistic. Relationships and societies in which behavior is very different with higher-ranking and lower-ranking individuals are called despotic, whereas those in which the influence of dominance on behavior is minimal are called egalitarian.[12] Rhesus females have nepotistic and despotic relationships with one another, and rhesus society as a whole can be best described as a highly nepotistic and despotic society.

Nepotism/individualism and despotism/egalitarianism are virtually universal dimensions of social relationships. Variation along these dimensions can be effectively used to characterize many animal and human societies. In nepotistic societies, individuals have strong preferences for their relatives as social partners (with the exception of sexual activities) and spend a lot of time with them. Most of their altruistic behavior is directed toward their kin, and kin are always helped against nonkin. Social interactions with nonrelatives are rare and are mostly competitive or antagonistic. Social business transactions in which altruistic behavior is traded for other services are occasionally exchanged with nonkin but are strictly regulated by tit-for-tat rules. Individuals' opportunities for social success in nepotistic societies depend in large part on the political power of their families. Individuals have limited opportunities to strike their own deals with unrelated individuals or groups, and social mobility across strata is strongly constrained by family pedigree. In nepotistic societies, destiny is set at birth. In contrast, in individualistic societies, social ties between adult individuals and their families of origin are maintained more loosely, and individuals can be socially attracted to nonkin and form all kinds of partnerships with them. Success is determined less by support from the family and more by individual characteristics, these characteristics being physical (such as strength or attractiveness), motivational (such as drive and ambition), or temperamental (such as a gregarious personality or resilience to adversity). Most importantly, individuals' social skills and their ability to form social bonds and establish alliances with others are the key to success in individualistic societies. In these societies, there is high mobility of individuals across social strata, and the structure of political power is far less stable and crystallized than in nepotistic societies.

Variation along the despotic/egalitarian dimension has to do with power, and in particular, with how differences in dominance between individuals affect their relationships. In despotic societies, there are strong and stable dominance hierarchies among individuals, differences in dominance are associated with large differences in power (such as in freedom of action, possession of material resources, or influence on other individuals), and they affect virtually every aspect of social life. Interactions between dominants and subordinates are very asymmetrical and rarely reciprocal. Dominant individuals assert their power and privileges in each and every circumstance and are rarely friendly toward subordinates. Instead, dominants exploit subordinates and control their behavior with any form of intimidation, oppression, or manipulation at their disposal. Tension and struggles for power between individuals of different rank are constant, but mobility across dominance ranks is allowed only through particular mechanisms or rules (for example, seniority). In contrast, in egalitarian societies, dominance hierarchies may be absent or their influence on social life may be negligible. Social interactions are based much more on cooperation and sharing, or negotiation and bargaining, than on oppression and exploitation. If dominance relationships exist, they are transient, reversible, and not associated with large differences in power between individuals. Dominant individuals are tolerant of subordinates and let them get a share of the pie almost as large as their own.

Nepotism and Despotism in Human Societies

Among all of the monkeys and apes, rhesus macaques have the quintessential despotic and nepotistic society, mainly because of the despotic and nepotistic relationships rhesus females have with one another. Humans can be quite flexible and adjust to their circumstances, but when all the outer layers of individualism and egalitarianism are peeled off, they have a despotic and nepotistic core that is not unlike that of rhesus macaques. The predominant lifestyle of a human society can affect the extent to which kinship and dominance affect social relationships. Individual geographic mobility, in particular, is a key factor that affects whether a social system is individualistic or nepotistic.[13] Groups of people with a nomadic lifestyle generally have few opportunities to form stable and long-term alliances with relatives or other individuals, and therefore their societies tend to be individualistic rather than nepotistic. Nomadic pastoralists who travel around over large areas with their cattle and are accompanied only by a few close family members tend to have individualistic societies of this kind. In contrast, groups of people with a sedentary lifestyle, who spend their lives surrounded by their family members and engage in activities that require the participation of many people, tend to form nepotistic social organizations. Agricultural and industrialized societies with low individual mobility and low rates of emigration tend to have nepotistic societies of this kind.

A good example of the relationship between low individual mobility and nepotistic social dynamics is provided by the Italian academic system. In Italy, most students go to college in the city in which they were born and end up working and spending the rest of their lives only a few blocks away from where their parents live. As a result, parents can influence and support the careers of their children much as rhesus macaque mothers do for their daughters. In contrast, for well-known historical and geographic reasons, American society is characterized by high mobility of individuals, with high rates of immigration from other countries and high rates of relocation within the country. American society as a whole is less nepotistic than Italian society, and the American academic system is less nepotistic than the Italian system. Most American students attend college in places far away from where their parents live, and those who pursue an academic career are always encouraged to move around to make sure their education is heterogeneous and diversified. Between starting out as an undergraduate student and becoming a professor, an American is likely to change geographic and intellectual environments at least four or five times, and all of this mobility discourages nepotistic mechanisms of career advancement. Nevertheless, family pedigree always matters, and nepotism does occur in America, especially in the upper strata of political and economic power. After all, it is no coincidence that two recent American presidents happen to have the same name.

Variation in human societies along the dimension of despotism and egalitarianism is accounted for not so much by patterns of individual mobility and geographic dispersal as by availability of resources (for example, wealth) and the extent to which those resources can be effectively monopolized by some individuals. Anthropologists have long noticed that contemporary societies of hunter-gatherers have a relatively egalitarian social organization.[14] It is possible that both hunting and gathering require the collaboration of many individuals, and that these kinds of subsistence activities encourage egalitarian social relationships. Another possibility is that a lifestyle based on these subsistence activities does not allow individuals to accumulate disproportionate amounts of wealth relative to others, but results in most individuals having similar access to and possession of resources. This, in turn, might encourage egalitarianism. Agricultural and industrialized societies, on the other hand, in which goods can be produced in great quantities and traded or sold to others, are more likely to produce large disparities in wealth and in political power between those individuals who are successful and those who are not. Differences in wealth and political power may favor the formation of dominance hierarchies and the organization of social relationships in relation to differences in dominance. Possession and exploitation of land on a permanent basis may also encourage territorial disputes and wars between groups. Groups with military power may use that power not only against other groups, but also within their own group, resulting in highly hierarchical and despotic social dynamics.

Nepotism itself may encourage despotism because nepotism inevitably leads to the creation of long-lasting differences in power and dominance between individuals or groups. In individualistic human societies, people count mostly on themselves, and although some end up being more successful than others, opportunities for success are, at least initially, evenly distributed among individuals. In nepotistic societies, on the other hand, individuals with more relatives are likely to have more power than those with fewer relatives. Nepotistic behavior amplifies and exacerbates differences in power and renders those differences very stable over time. Whereas in individualistic societies the balance of power can be modified with bargaining and negotiation, in highly nepotistic societies the structure of power is so entrenched that it takes a revolution to change it drastically.

Human Nature and Rhesus Nature

Variations in lifestyle, including patterns of individual mobility and dispersal, types of subsistence activity, and opportunities to accumulate and monopolize resources, are only some of the factors that result in variation among human societies along the dimensions of nepotism/individualism and despotism/egalitarianism. There are many other possible factors, including, for example, mating (marriage) systems and systems of parental care. Finally, superimposed on all these factors, there are cultural influences—for example, historical or religious traditions—that, depending on the part of the world, might encourage individuals to be nepotistic or individualistic, despotic or egalitarian. According to some social scientists, all of this variation in lifestyle, social organization, and behavior defines humans as a species and makes it impossible to know what human nature really is because there is no such thing.[15] I disagree, and so do many others.[16]

People from different parts of the world make themselves look very different with the different cultural clothes they wear. When they take their clothes off, however, their bodies look the same. Rhesus macaques don't wear any clothes, and their behavior and social organization look pretty much the same no matter what part of the world they live in. They seem to have a strong predisposition to form highly despotic and nepotistic societies wherever they are. To know human nature and understand whether people have a biological predisposition to form particular types of social organizations, we have to strip them of their cultural clothes and take a peek at their naked bodies.

Life in jails or prison camps around the world might tell us something about how people organize themselves and what kinds of social relationships they form with one another when all the historical, geographic, and cultural factors that normally affect human behavior are thrown out the window. When convicted criminals enter a prison, they leave all their clothes and personal belongings at the door—not just their shirts, pants, and wallets, but their cultural clothes as well. They leave behind their family history and upbringing, their education and religious beliefs, their political power and material possessions, and all of their past accomplishments and plans for the future. In other words, they are stripped of their cultural identities, which normally define the way they think about themselves, the way others view them, and their place in society. The clock of their lives gets reset and they must start from scratch, but without all the social and cultural scaffolding that normally surrounds and supports life outside the prison. All they get is a number and a uniform.

Although I am no expert on this subject, accounts of social life in jails and prison camps around the world tend to be quite similar regardless of geographic location, historical period, or the prisoners' ethnic, cultural, or religious backgrounds. Prisoners find themselves competing with one another for whatever resources are available, and competition for resources and power results in the establishment of dominance relationships between them. Survival and acquisition of power in a jail or prison camp depend on protection and support from others, so prisoners also form social bonds and cooperative alliances with some other prisoners. In effect, they re-create small kin groups and engage in nepotistic behavior the way they would with their family members. The result is a highly despotic and nepotistic social organization, not unlike that of rhesus macaques. A powerful and dramatic description of this process can be found in Primo Levi's account of life in a Nazi concentration camp, *If This Is a Man.*[17] In this book, Levi provides not only a description of the cruelty and harm inflicted by the Nazis on their prisoners, but also a courageous and honest account of how the prisoners treated one another. Levi was haunted his whole life by what he saw and eventually committed suicide.

Social life within military organizations around the world also seems to be organized according to rigid despotic and nepotistic rules of behavior. Like imprisonment, entry into a military organization, whether voluntary or coerced, involves the loss of one's cultural identity and previously accumulated cultural baggage. When I was drafted into the military, I immediately realized that my upbringing, education, previous accomplishments, and plans for the future no longer mattered. I felt naked and vulnerable. It didn't take me long to figure out that I found myself in a highly despotic and nepotistic social system and that I was right at the bottom of the hierarchy with no friends and relatives around me. All of a sudden I had been turned into a rhesus macaque who had just transferred into a new group.

Armies around the world and throughout human history seem

to have a very similar hierarchical structure, and that structure seems to encourage similar social relationships among the soldiers. I've never heard of an army without a strict dominance hierarchy, in which all soldiers have egalitarian relationships with one another and every relationship is negotiated on a one-to-one basis, depending on the circumstances and without the influence of other individuals. I suspect that armies have the structure they do simply because that structure is what works best, given what armies need to accomplish. The function of armies is to fight with other armies and kill enemies. Armies need to be internally stable, cohesive, and capable of well-coordinated action. A strong hierarchical structure makes an army a very stable and cohesive social organization and facilitates the decision-making process involved in confrontations with the enemy. Suppression or minimization of positive emotions such as love and compassion between soldiers and maintenance of a simple but effective system of communication also facilitate the maintenance of stability and cohesiveness. Finally, social opportunism and nepotism within the army may encourage the soldiers to engage in ingroup versus outgroup thinking and behavior, and these are crucial for their motivation to fight. Soldiers who don't care about ingroup versus outgroup differences and tend to treat everybody the same way would have little or no motivation to kill others or to risk their lives in combat. Soldiers who are highly motivated to protect their own selfish interests and those of their buddies, however, make perfect war machines. So a despotic and nepotistic organization makes an army a perfect war machine. The social structure of an army is nicely adapted to the function the army is meant to serve, just as the anatomy and shape of a bird's beak is adapted to its function, for example, of cracking hard seeds.[18] Rhesus macaque groups seem to be structured and to function according to the same rules.

I suspect that people, and especially men, have a biological predisposition to organize themselves into highly despotic and nepotistic societies when all the environmental, historical, and cultural influences that normally affect their behavior are suppressed. This predisposition may be the result of our evolutionary history. People do not behave the way they do in jails or prison camps because humans spent most of their evolutionary history in these conditions of imprisonment. Instead, people probably spent much of their evolutionary history living in military-like social organizations and fighting with other groups of people.

Of course, over the last few thousand years, many people have come to appreciate the benefits of a peaceful lifestyle and have learned to wear all kinds of fancy cultural clothes. To adjust to living peacefully in societies with thousands of other unrelated and unfamiliar human beings, humans might also have developed biological predispositions to be nice, not only to family members, but also to strangers. We might also have developed an acquired biological taste for egalitarianism and the moral principles that support it. However, this is all relatively new baggage that we carry in our brains, just like the human male's predisposition to invest in his offspring.[19] Our military-style mentality and the despotic and nepotistic tendencies that come along with it still permeate many aspects of our social lives, and when we take our cultural clothes off, we go right back to it. We probably have a history of violence as a species, and it comes back to haunt us all the time. Our psychological and social predispositions for despotism and nepotism might go as far back in time as the ancestor rhesus macaques and people had in common. Here is a hypothetical scenario of how things came to be the way they are.

Rhesus Macaques and Humans: A Success Story

Whenever there is a change in the environment—for example, in the average temperature or the type of available food—those individuals who have the genes to exploit the new environment are more likely to survive and reproduce than those without those genes. Because some individuals and their genes move forward and others are left behind from every generation to the next, there is always a good fit between the environment and the organisms that live in it. In other words, organisms are always well adapted to their current environment.

For many animal species, the main selective pressures driving evolution are the things that threaten their survival and ability to reproduce, such as the availability of food and the risk of predation. Different species adapt to problems of feeding and predator avoidance in different ways. Some species become highly specialized in what they eat and how they eat it. They can become very well adapted to their feeding niche, but if the environment changes suddenly and dramatically, these specialists don't do well anymore. Other species, instead, become ecologically flexible generalists. They can eat a number of different foods and survive and reproduce well in a variety of climatic conditions. In an environment with many repeated drastic changes, these species do very well. Both rhesus macaques and humans took the generalist approach, and it worked well for them. There are also different solutions to the predation problem. Some species become small, solitary, and nocturnal to avoid predators. Others increase in body size or form social groups to protect themselves. Very small and very large body sizes have their evolutionary advantages and disadvantages, their pros and cons. Rhesus macaques and people settled for medium body size and life in large groups, and this choice worked out well for them.

Once life in large groups evolved, something interesting probably began to happen. Cooperation and competition between individuals and between groups gradually became more important in determining survival and reproduction than finding food or avoiding predators. Individuals and groups that were socially successful had much higher chances of surviving and reproducing than socially unsuccessful individuals and groups and passed on more copies of their genes to the next generation. When this happened, the social environment became the main selective pressure for the evolution of psychological and behavioral traits.[20] For example, as groups became larger and more opportunities for complex patterns of cooperation and competition both within and between groups arose, pressures mounted for an increase in Machiavellian intelligence, and perhaps for increased intelligence in general.

So what rhesus macaques and people may have in common is that their psychological and behavioral dispositions have been shaped by selective pressures stemming from cooperation and competition between individuals and groups to a larger extent than many other animal species, including many other primates. The main claim to fame of some animal species is how they eat their food or how they avoid predators. Cows have a stomach with four chambers that allows them to eat grass all day. Birds fly. In contrast, the most striking adaptations of rhesus macaques and people are social and cognitive. We are socially smart. We are Machiavellian.

Fights between groups have probably been frequent and intense throughout the evolutionary history of rhesus macaques. The same is probably true of wars between groups throughout hominid and human history. This kind of competition between groups promotes the need for cooperation between individuals of the same group. These pressures for cooperation and competition may have resulted not only in social dispositions toward despotism and nepotism, but also in particular temperamental characteristics. Both rhesus macaques and people have a generally gregarious and aggressive temperament; that is, they like to spend time with individuals they like, but they also fight with them a lot. In rhesus macaques, mothers squabble with their offspring all the time, siblings fight with one another, older females fight with all of their female relatives, and adult males fight with everybody else. Humans are not that different. Squabbles between family members are frequent and intense, and most murders involve people who know each other well. We need other people, we love them, but we can't help fighting with them.

Rhesus macaques and people are also similar in the way they react to novelty and threats. Individuals of both species are generally curious about their surroundings, and this characteristic facilitates their tendency to explore, colonize, and adapt to novel environments, including novel foods. Other species of primates, including other macaques, are easily put off by novelty. If you put them in a novel environment, they freeze or withdraw instead of exploring.[21] Rhesus macaques also have the tendency to engage strangers and react aggressively to potential threats from them instead of avoiding and withdrawing. For example, whereas many monkeys and apes withdraw and show signs of fear if approached by people they don't know, rhesus macaques instead approach and threaten a stranger who is looking at them. They show fear too, but their fear doesn't keep them from investigating and threatening the stranger, and most importantly, from calling other rhesus macaques to come to their aid. Rhesus macaques always engage the enemy in a confrontation and don't withdraw unless you give them a good reason to do so.

Finally, rhesus macaques don't seem to waste their energy on positive emotions and feelings, such as love and compassion, or in sophisticated forms of communication that might get in the way of the effective functioning of their societies. These temperamental traits and social dispositions have probably played an important role in the ecological and social success of rhesus macaques. Human beings share some of these traits, and they may be one of the reasons for our success as colonizers and conquerors on this planet. These temperamental traits are characteristic of weed species, and in the case of the rhesus and human weeds, they come in a package that contains nepotism and despotism as well.

So even though rhesus macaques have taken the route of female bondedness and female dominance while humans have become more of a male-centered and male-dominated species, the evolutionary journey of these two species has been remarkably similar and has touched many of the same destinations. Both rhesus macaques and humans exemplify the success story of a medium-sized and average-looking primate with a curious, gregarious, and aggressive temperament. Neither species has any fancy specializations for feeding or fighting; both are highly adaptable to novel environments and circumstances, and both specialize in social life and social intelligence. Like many other animal species, rhesus macaques and humans opted for life in large social groups to increase their safety and competitive ability, but by adopting a highly despotic and nepotistic social organization, they turned these groups into effective war machines as well as engines for the production of Machiavellian brains. Once the production of Machiavellian human brains was set in motion, their size and complexity continued to increase exponentially, and we continued our evolutionary journey to cognitive destinations beyond the reach of even the most ambitious rhesus macaque traveler. With our big brains came many new intellectual skills that now set us apart from all the other animals. That, however, happened later—much later.

Rhesus macaques in many ways embody some of the worst aspects of human nature, but I suspect that the hominids and early humans from which we evolved weren't "nice" animals. For most of our evolutionary history we probably acted a lot like rhesus macaques, and we still do, in our everyday lives. Rhesus macaques might tell us how our journey of cognitive and social evolution got started and why. Our history of violence and our Macachiavellian intelligence are not necessarily things we can be proud of, but they may be the secret of our success. If they have driven the evolution of our large brains, they may also be what has made us capable of many other beautiful things, including our ability to engage in noble and rewarding spiritual and intellectual activities and our love and compassion for other people.

CHAPTER 1

1. Niccolò Machiavelli (1469–1527) first announced the completion of his book in a letter dated December 10, 1513. Machiavelli originally gave his book a title in Latin, *De Principatibus*, but a few years later it was published with the Italian title of *Il Principe*. An English translation is available in a paperback edition (Machiavelli 1984). Machiavelli dedicated his book to Lorenzo II de' Medici, who did not like it much. The prince referred to by Machiavelli is Cesare Borgia (1475–1507), a skillful and ruthless politician and warrior who conquered the city of Urbino and other territories in central Italy before he died at the age of thirty-two.
2. This statement was written in 1838 by Darwin in his Notebook M, *Meta physics on Morals and Speculations on Expression* (in Barrett et al. 1987).

CHAPTER 2

1. Tool use is more common in chimpanzees and other great apes than in monkeys, with the exception of capuchin monkeys, a group of South American monkeys belonging to the genus *Cebus*, which use rocks, sticks, and other tools to obtain or process their food. For information about the distribution and evolution of tool use in nonhuman primates, see van Schaik et al. 1999.
2. Sir David Attenborough (1926–) is a well-known host of BBC nature documentaries. In these documentaries, he is often seen in the field describing rare and complex aspects of animal behavior while the animals are performing the behavior in question right behind him.
3. Recent reclassifications of macaques on the basis of genetic data have largely confirmed the earlier taxonomy proposed by Fooden (1980) on the basis of anatomical and morphological characteristics. Some disagreement still exists, however, concerning the classification of a few macaque species.
4. See www.LanguageMonitor.com.
5. In *Parade Magazine*, August 21, 2005.

6. Comprehensive information on the evolutionary history, ecology, and behavior of macaques can be found in three edited volumes (Lindburg 1980; Fa and Lindburg 1996; Thierry et al. 2004) and several review articles (e.g., Fa 1989; Abegg and Thierry 2002).
7. Barbary macaques (*Macaca sylvanus*) are in a group of their own. The other species are grouped into three clusters: the *silenus* group (*Macaca brunnescens, M. hecki, M. maurus, M. nemestrina, M. nigra, M. nigrescens, M. ochreata, M. silenus,* and *M. tonkeana*), the *radiata* group (*Macaca arctoides, M. assamensis, M. radiata, M. sinica,* and *M. thibetana*), and the *fascicularis* group (*Macaca cyclopis, M. fascicularis, M. fuscata,* and *M. mulatta*) (see Delson 1980).
8. Other researchers dispute the claim by Southwick and his colleagues (1996) that rhesus macaques are the primate species with the second broadest geographic distribution after humans (Dr. Alexander Harcourt, personal communication, 2005).
9. The distinction between weed and nonweed macaques was first made by Richard et al. (1989).
10. The rhesus macaque and a closely related species, the long-tailed macaque, often benefit from these parties. One of these parties for long-tailed macaques in Thailand is shown in plate 1.
11. Detailed information on the island of Cayo Santiago and its population of rhesus macaques is provided in an edited volume (Rawlins and Kessler 1986a). A detailed account of how Carpenter established the colony is provided in a chapter of that volume by Rawlins and Kessler (1986b).
12. The story of the establishment of the hamadryas baboon colony at the London Zoo in 1925 is told by Zuckerman (1932).
13. Hartman had studied rhesus macaque reproduction in the laboratory (e.g., Hartman 1931). His statement about reproduction on Cayo Santiago is reported by Rawlins and Kessler (1986b, 15).
14. Many of these social deprivation studies were conducted at the University of Wisconsin by Harry Harlow and his students. Harlow and his students also demonstrated the importance of physical contact to other monkeys for their well-being and normal development. The story of this research is told in Harlow's biography (Blum 2002).

CHAPTER 3

1. William D. Hamilton (1936–2000) published two very important papers based on his Ph.D. dissertation work in the mid-1960s (Hamilton 1964a, 1964b), in which he showed with mathematical models that a gene for altruism can spread through a population if the cost to the altruist is lower than the benefit to the recipient devalued by a fraction representing the

genetic relatedness between actor and recipient (the coefficient of relatedness). Hamilton's papers have had a huge influence on research in behavioral biology and on our understanding of the evolution of altruism, but they were poorly understood when they were first published and were ignored for a decade.

2. Sigmund Freud (1856–1939) proposed his theory of sexual attraction during childhood in a collection of essays first published in 1905 (for a recent English translation in a paperback edition, see Freud 2000).
3. Edvard Westermarck (1862–1939) was a Finnish sociologist who brought attention to the finding that children who grow up together exhibit little or no mutual sexual attraction later in life (Westermarck 1921). One of the best-known demonstrations of the Westermarck effect involves the Israeli kibbutz system, in which children are raised in a common children's house, away from their parents; later in life, they show little or no sexual interest in the members of their own cohort.
4. See Colvin 1986.
5. For a discussion of differences in primate dispersal and their consequences for social organization, see Isbell 2004.
6. Wrangham 1979.
7. See Rodseth et al. 1991.
8. Rodseth et al. 1991.
9. Eugene Marais (1872–1936) was a journalist, lawyer, poet, and natural scientist from South Africa who, between 1905 and 1910, spent three years observing the behavior of a troop of chacma baboons in Northern Transvaal. The quotation in the text is from his book *My Friends the Baboons*, which was first translated into English in 1939, three years after his death (Marais 1939).
10. See Thierry 2000. Primatologists define reconciliation as the occurrence of an affiliative interaction (e.g., grooming) between two individuals within ten minutes after they've had a fight. To demonstrate the occurrence of reconciliation, they compare the frequency of occurrence of affiliation in the ten-minute period after a fight with the frequency of occurrence of affiliation between the same two individuals in a ten-minute period the day following the fight, at the same time.
11. Mother Teresa of Calcutta (1910–1997) was an Albanian Roman Catholic nun who spent most of her life attending to the physical and spiritual needs of the poor. She received the Nobel Peace Prize in 1979 and was beatified (i.e., declared a saint) by Pope John Paul II in 2003. See Hitchens 1997 for a skeptical look at the motives of Mother Teresa's altruism.
12. My analogy between software development and natural selection should not be interpreted as implying that natural selection is guided by a con-

scious agent. Natural selection is a blind process guided by no agent, conscious or unconscious.

13. See Wasser 1983; Maestripieri 2001a.
14. Nepotistic behavior is widespread among animals with limited cognitive abilities, such as social insects.
15. See Berman 2004.
16. For a discussion of self-referent phenotype matching (the "armpit effect"; Dawkins 1982) and other kin recognition mechanisms, see Sherman et al. 1997.
17. Widdig et al. 2001, 2006.

CHAPTER 4

1. Konrad Lorenz (1903–1989) is recognized for being, along with Dutch colleague Niko Tinbergen (a co-winner of the Nobel Prize in 1973 with Lorenz and Karl von Frisch), one of the founders of ethology, the scientific discipline that studies the biological bases of behavior. His theories about aggressive behavior were popularized in a book published in German in 1963 with the title *Das sogenannte Böse* [The So-Called Evil], which was later translated into English and published with the title *On Aggression* (Lorenz 1966).
2. See Frank et al. 1991.
3. The important implications of the invention of projectile weapons for human social evolution are discussed in Bingham 1999.
4. Golding 1954.
5. Hinde 1976.
6. Chance 1967.
7. In some cases, however, subordinates do take advantage of the absence of dominants from their group (e.g., Drea and Wallen 1999).
8. Morris 1996.
9. See Gouzoules et al. 1984.
10. For example, see Morgan 1978.
11. Most of this description of patterns of agonistic intervention by rhesus macaque males is based on unpublished data collected on Cayo Santiago.
12. Retaliation for aggression against one of the aggressor's relatives has been demonstrated in a closely related macaque species, the Japanese macaque (Aureli et al. 1992), but not in rhesus macaques.
13. This signal is, in most cases, the fear grin, or bared-teeth display, described in chapter 8 (see Maestripieri 1996).
14. For experimental evidence for the role of agonistic support in rank acquisition, see Chapais 1988.

15. For example, see Schulman and Chapais 1980.
16. See Sapolsky 2005.

CHAPTER 5

1. For example, see Brewer 1999.
2. For further information about the first encounter between the New Guinea highlanders and the outside world, see Connolly and Anderson 1987.
3. See Gallup 1998.
4. This pattern was observed on Cayo Santiago (Chepko-Sade and Sade 1979), but things may be different in the wild (Melnick and Kidd 1983).
5. See Ehardt and Bernstein 1986 and Gygax et al. 1997.

CHAPTER 6

1. Beach 1947; Wallen 1995.
2. Wallen 1990.
3. Zuckerman 1932.
4. For example, see Michael and Welegalla 1968.
5. Michael et al. 1971; see Goldfoot 1981 for a critique.
6. Carpenter 1942.
7. Gordon 1981; Wallen 1982.
8. One study, however, showed that adult male baboons intervene on behalf of their juvenile offspring in agonistic disputes with other individuals, thus suggesting that they somehow discriminate their offspring from other juveniles (Buchan et al. 2003).
9. This, of course, may be true in a very hypothetical and context-free situation, which would never occur in real life.
10. Darwin 1872.
11. The mating systems in which this occurs are called "leks" (see Höglund and Alatalo 1995).
12. See Berard 1990; Widdig et al. 2003.
13. See Berard et al. 1994.
14. Bercovitch 1997.
15. See also Berard et al. 1994.
16. Gordon et al. 1976.
17. Zehr et al. 1998.
18. See Berard 1990; Manson 1995.
19. But rhesus females often exhibit an aversion to mating with their male relatives (Manson and Perry 1993).
20. See Alatalo and Ratti 1995.

21. See Gangestad and Thornhill 1997; Roney et al. 2006.
22. See Adams et al. 1978 for evidence that female-initiated sexual activity rises around the time of ovulation.
23. See van Schaik and Janson 2000.
24. The best evidence that females are attracted to males who can provide protection from infanticide comes from primate species other than rhesus macaques (see van Schaik and Janson 2000). Male infanticide is not well documented in rhesus macaques (but see Camperio Ciani 1984).
25. vom Saal 1985.
26. But see Soltis and McElreath 2001.
27. This study involved not rhesus macaques, but pigtail macaques; see Gust et al. 1996.
28. See Birkhead and Parker 1997.

CHAPTER 7

1. Parker 1996.
2. Maestripieri 2005a.
3. Trivers 1972.
4. For example, see Maestripieri and Pelka 2002.
5. Leveroni and Berenbaum 1988.
6. Hassett et al. 2004. See also Alexander and Hines 2002 for a similar study in vervet monkeys.
7. The best evidence that infant handling helps young females become better mothers comes from a study of vervet monkeys (Fairbanks 1990).
8. See Schino and Troisi 2004.
9. For example, see Warren and Brooks-Gunn 1989.
10. For example, see Maestripieri 1991.
11. Maestripieri 2001b.
12. Clever Hans was a horse that was seemingly able to perform arithmetic tasks. In reality, the correct answers he gave to the questions he was asked were guided by involuntary cues in the face or body of his trainer (Pfungst 1911).
13. Maestripieri 2001b.
14. For example, see Ordog et al. 1998.
15. Simpson et al. 1981.
16. Maestripieri 2002.
17. Trivers 1974.
18. See Soltis 2004; Maestripieri and Durante 2004.
19. See Haskell 1994 for evidence that begging increases the risk of predation.
20. Berman et al. 1994.
21. Maestripieri 2004.

22. Brain 1992.
23. Maestripieri 1994.
24. Whitham et al. 2007.

CHAPTER 8

1. The expression "representational communication" means different things to different people. Some researchers define as representational only signals that refer to other individuals or objects in the external environment (e.g., Cheney and Seyfarth 1990). According to others, signals that convey information about the signaler's emotions, motivation, or future behavior can also be interpreted as representational (e.g., Owren et al. 2003). I adopt the latter use of the word.
2. For example, see Fernald 1992.
3. Owren et al. 2003.
4. For example, see Burling 1993.
5. For example, see Cheney and Seyfarth 1990.
6. Savage-Rumbaugh and Lewin 1994.
7. Heyes 1998; Tomasello et al. 2005; Cheney and Seyfarth 2007.
8. Blubber and Mr. T were the names of two males in a group of pigtail macaques at the Yerkes National Primate Research Center in the mid-1990s. Social interactions between adult males in pigtail macaques are very similar to those in rhesus macaques.
9. Maestripieri and Wallen 1997.
10. Maestripieri 2005b.
11. For example, see Aureli and van Schaik 1991.
12. See Maestripieri 1996 for a critique of this interpretation.
13. Maestripieri 1996, 1999.
14. Tomasello and Call 1997.
15. Maestripieri 1999.
16. Van Hooff 1972.
17. For example, see Fromhage and Schneider 2005.
18. Vahed 1998.
19. Whitham et al. 2007.
20. Maestripieri 1995.
21. Dunbar 1998.
22. Maestripieri 2005b.

CHAPTER 9

1. See Harcourt and Stewart 2007.
2. See Maestripieri 2005c for a discussion of King Kong's behavior from a primatological perspective.

3. Jerison 1973.
4. See Humphrey 1976; Byrne and Whiten 1988.
5. Sawaguchi 1992; Dunbar 1993.
6. See Wrangham 1979; Lindenfors 2004.
7. See Altmann 1990; van Schaik 1989.
8. Lindenfors 2005.
9. Rodseth et al. 1991.
10. Wrangham 1979; van Schaik 1989.
11. For example, see Packer 1977.
12. For a discussion of nepotistic versus individualistic and despotic versus egalitarian societies in animals, especially nonhuman primates, see Vehrencamp 1983 and van Schaik 1989.
13. For a discussion of nepotistic versus individualistic and despotic versus egalitarian human societies, see Knauft 1991 and Boehm 1999.
14. Knauft 1991; Boehm 1999.
15. Durkheim 1895/1962; Geertz 1973.
16. For example, Tooby and Cosmides 1992.
17. Primo Levi (1919–1987) wrote *Se Questo e' un Uomo* (If This Is a Man) in 1947. For an English translation in a paperback edition, see Levi 1979.
18. See Weiner 1995.
19. Although humans probably spent a significant portion of their early evolutionary history living in social conditions similar to those of contemporary hunter-gatherer societies, what I argue here is that many significant aspects of human nature—many of our psychological and behavioral predispositions—have a much more ancient evolutionary history, and may go back to the ancestors we share with Old World monkeys and apes. Many similarities between rhesus macaques and humans, however, may reflect adaptation to similar environments and the fact that both are "weed" species, rather than common ancestry. This possibility is discussed in the next section.
20. See Alexander 1974.
21. See Clarke and Boinski 1995.

References

Abegg, C., and B. Thierry. 2002. Macaque evolution and dispersal in insular south-east Asia. *Biological Journal of the Linnean Society* 75:555–76.

Adams, D. B., A. Ross Gold, and A. D. Burt. 1978. Rise in female-initiated sexual activity at ovulation and its suppression by oral contraceptives. *New England Journal of Medicine* 299:1145–50.

Alatalo, R. V., and O. Ratti. 1995. Sexy son hypothesis: Controversial once more. *Trends in Ecology and Evolution* 10:52–53.

Alexander, G. M., and M. Hines. 2002. Sex differences in response to children's toys in nonhuman primates (*Cercopithecus aethiops sabaeus*). *Evolution and Human Behavior* 23:467–79.

Alexander, R. D. 1974. The evolution of social behaviour. *Annual Review of Ecology and Systematics* 5:325–83.

Altmann, J. 1990. Primate males go where the females are. *Animal Behaviour* 39:193–95.

Aureli, F., R. Cozzolino, C. Cordischi, and S. Scucchi. 1992. Kin-oriented redirection among Japanese macaques: An expression of a revenge system? *Animal Behaviour* 44:283–91.

Aureli, F., and C. P. van Schaik. 1991. Post-conflict behaviour in long-tailed macaques: I. The social events. *Ethology* 89:89–100.

Barrett, P. H., P. J. Gautrey, S. Herbert, D. Kohn, and S. Smith, eds. 1987. *Charles Darwin's Notebooks, 1836–1844: Geology, Transmutation of Species, Metaphysical Enquiries.* London: British Museum (Natural History); Cambridge: Cambridge University Press.

Beach, F. A. 1947. Evolutionary changes in the physiological control of mating behavior in mammals. *Psychological Review* 54:297–315.

Berard, J. D. 1990. Life history patterns of male rhesus macaques on Cayo Santiago. Ph.D. dissertation, University of Oregon.

Berard, J. D., P. Nurnberg, J. T. Epplen, and J. Schmidtke. 1994. Alternative reproductive tactics and reproductive success in male rhesus macaques. *Behaviour* 129:177–201.

Bercovitch, F. B. 1997. Reproductive strategies of rhesus macaques. *Primates* 38:247–63.

Berman, C. M. 2004. Developmental aspects of kin bias in behavior. In *Kinship and Behavior in Primates,* ed. B. Chapais and C. M. Berman, 317–46. Oxford: Oxford University Press.

Berman, C. M., K. L. R. Rasmussen, and S. J. Suomi. 1994. Responses of free-ranging rhesus monkeys to a natural form of social separation. I. Parallels with mother-infant separation in captivity. *Child Development* 65:1028–41.

Bingham, P. M. 1999. Human uniqueness: A general theory. *Quarterly Review of Biology* 74:133–69.

Birkhead, T. R., and G. A. Parker. 1997. Sperm competition and mating systems. In *Behavioural Ecology: An Evolutionary Approach,* 4th ed., ed. J. R. Krebs and N. B. Davies, 121–45. Oxford: Blackwell.

Blum, D. 2002. *Love at Goon Park: Harry Harlow and the Science of Affection.* Cambridge, MA: Perseus Publishing.

Boehm, C. 1999. *Hierarchy in the Forest: The Evolution of Egalitarian Behavior.* Cambridge, MA: Harvard University Press.

Brain, C. 1992. Deaths in a desert baboon troop. *International Journal of Primatology* 13:593–99.

Brewer, M. B. 1999. The psychology of prejudice: Ingroup love or outgroup hate? *Journal of Social Issues* 55:429–44.

Buchan, J. C., S. C. Alberts, J. B. Silk, and J. Altmann. 2003. True paternal care in a multi-male primate society. *Nature* 425:179–81.

Burling, R. 1993. Primate calls, human language, and nonverbal communication. *Current Anthropology* 34:25–53.

Byrne, R., and A. Whiten, eds. 1988. *Machiavellian Intelligence: Social Expertise and the Evolution of Intellect in Monkeys, Apes, and Humans.* Oxford: Oxford University Press.

Camperio Ciani, A. 1984. A case of infanticide in a free ranging group of rhesus monkeys (*Macaca mulatta*) in the Jakoo forest, Simla (India). *Primates* 25:372–77.

Carpenter, C. R. 1942. Sexual behavior of free-ranging rhesus monkeys (*Macaca mulatta*). I. Specimens, procedures, and behavioral characteristics of estrus. *Journal of Comparative Psychology* 33:113–42.

Chance, M. R. A. 1967. Attention structure as the basis of primate rank orders. *Man* 2:503–18.

Chapais, B. 1988. Experimental matrilineal inheritance of rank in female Japanese macaques. *Animal Behaviour* 36:1025–37.

Cheney, D. L., and R. M. Seyfarth. 1990. *How Monkeys See the World: Inside the Mind of Another Species.* Chicago: University of Chicago Press.

———. 2007. *Baboon Metaphysics*. Chicago: University of Chicago Press.

Chepko-Sade, B. D., and D. S. Sade. 1979. Patterns of group splitting within matrilineal kinship groups. *Behavioral Ecology and Sociobiology* 5:67–86.

Clarke, A. S., and S. Boinski. 1995. Temperament in nonhuman primates. *American Journal of Primatology* 37:103–25.

Colvin, J. D. 1986. Proximate causes of male emigration at puberty in rhesus monkeys. In *The Cayo Santiago Macaques: History, Behavior, and Biology*, ed. R. G. Rawlins and M. J. Kessler, 131–57. Albany: SUNY Press.

Connolly, B., and R. Anderson. 1987. *First Contact: New Guinea's Highlanders Encounter the Outside World*. New York: Viking Penguin.

Darwin, C. 1872. *The Descent of Man, and Selection in Relation to Sex*. London: Murray.

Dawkins, R. 1982. *The Extended Phenotype*. San Francisco: Freeman.

Delson, E. 1980. Fossil macaques, phyletic relationships and a scenario of deployment. In *The Macaques: Studies in Ecology, Behavior, and Evolution*, ed. J. E. Fa and D. G. Lindburg, 10–30. New York: Van Nostrand Reinhold.

Drea, C. M., and K. Wallen. 1999. Low-status monkeys "play dumb" when learning in mixed social groups. *Proceedings of the National Academy of Sciences USA* 26:12965–69.

Dunbar, R. I. M. 1993. Coevolution of neocortical size, group size, and language in humans. *Behavioral and Brain Sciences* 16:681–94.

———. 1998. *Grooming, Gossip, and the Evolution of Language*. Cambridge, MA: Harvard University Press.

Durkheim, E. 1895/1962. *The Rules of Sociological Method*. Glencoe, IL: Free Press.

Ehardt, C. L., and I. S. Bernstein. 1986. Matrilineal overthrows in rhesus monkey groups. *International Journal of Primatology* 7:157–81.

Fa, J. E. 1989. The genus *Macaca*: A review of taxonomy and evolution. *Mammal Reviews* 19:45–81.

Fa, J. E., and D. G. Lindburg, eds. 1996. *Evolution and Ecology of Macaque Societies*. Cambridge: Cambridge University Press.

Fairbanks, L. A. 1990. Reciprocal benefits of allomothering for female vervet monkeys. *Animal Behaviour* 40:553–62.

Fernald, A. 1992. Human maternal vocalizations to infants as biologically relevant signals: An evolutionary perspective. In *The Adapted Mind: Evolutionary Psychology and the Generation of Culture*, ed. J. H. Barkow, L. Cosmides, and J. Tooby, 267–88. Oxford: Oxford University Press.

Fooden, J. 1980. Classification and distribution of living macaques (*Macaca* Lacepede, 1799). In *The Macaques: Studies in Ecology, Behavior, and Evolution*, ed. D. G. Lindburg, 1–9. New York: Van Nostrand Reinhold.

Frank, L. G., S. E. Glickman, and P. Licht. 1991. Fatal sibling aggression, precocial development, and androgens in neonatal spotted hyenas. *Science* 252:702–4.

Freud, S. 2000. *Three Essays on the Theory of Sexuality.* New York: Basic Books.

Fromhage, L., and J. M. Schneider. 2005. Safer sex with feeding females: Sexual conflict in a cannibalistic spider. *Behavioral Ecology* 16:377–82.

Gallup, G. G. Jr. 1998. Self-awareness and the evolution of social intelligence. *Behavioural Processes* 42:239–47.

Gangestad, S. W., and R. Thornhill. 1997. The evolutionary psychology of extra-pair sex: The role of fluctuating asymmetry. *Evolution and Human Behavior* 18:69–88.

Geertz, C. 1973. *The Interpretation of Cultures.* New York: Basic Books.

Goldfoot, D. A. 1981. Olfaction, sexual behavior, and the pheromone hypothesis in rhesus monkeys: A critique. *American Zoologist* 21:153–64.

Golding, W. 1954. *Lord of the Flies.* New York: Berkley Publishing Group.

Gordon, T. P. 1981. Reproductive behavior in the rhesus monkey: Social and endocrine variables. *American Zoologist* 21:185–95.

Gordon, T. P., R. M. Rose, and I. S. Bernstein. 1976. Seasonal rhythm in plasma testosterone levels in the rhesus monkey (*Macaca mulatta*): A three-year study. *Hormones and Behavior* 7:229–43.

Gouzoules, S., H. Gouzoules, and P. Marler. 1984. Rhesus monkey (*Macaca mulatta*) screams: Representational signalling in the recruitment of agonistic aid. *Animal Behaviour* 37:182–93.

Gust, D. A., T. P. Gordon, W. F. Gergits, N. J. Casna, K. G. Gould, and H. M. McClure. 1996. Male dominance rank and offspring-initiated behaviors were not predictors of paternity in a captive group of pigtail macaques (*Macaca nemestrina*). *Primates* 37:271–78.

Gygax, L., N. Harley, and H. Kummer. 1997. A matrilineal overthrow with destructive aggression in *Macaca fascicularis*. *Primates* 38:149–58.

Hamilton, W. D. 1964a. The genetical evolution of social behaviour. I. *Journal of Theoretical Biology* 7:1–16.

———. 1964b. The genetical evolution of social behaviour. II. *Journal of Theoretical Biology* 7:17–52.

Harcourt, A. H., and K. J. Stewart. 2007. *Gorilla Society: Conflict, Compromise, and Cooperation between the Sexes.* Chicago: University of Chicago Press.

Hartman, G. 1931. The breeding season in monkeys, with special reference to *Pithecus (Macaca) rhesus*. *Journal of Mammalogy* 12:129–42.

Haskell, D. G. 1994. Experimental evidence that nestling begging behaviour incurs a cost due to nest predation. *Proceedings of the Royal Society of London* B 257:161–64.

Hassett, J. M., E. R. Siebert, and K. Wallen. 2004. Sexually differentiated toy preferences in rhesus monkeys. *Hormones and Behavior* 46:91.

Heyes, C. M. 1998. Theory of mind in nonhuman primates. *Behavioral and Brain Sciences* 21:101–48.

Hinde, R. A. 1976. Interactions, relationships, and social structure. *Man* 11:1–17.

Hitchens, C. 1997. *The Missionary Position: Mother Teresa in Theory and Practice*. New York: Verso Publishing.

Hoglund, J., and R. V. Alatalo. 1995. *Leks*. Princeton, NJ: Princeton University Press.

Humphrey, N. 1976. The social function of intellect. In *Growing Points in Ethology*, ed. P. P. G. Bateson and R. A. Hinde, 303–17. Cambridge: Cambridge University Press.

Isbell, L. A. 2004. Is there no place like home? Ecological bases of dispersal in primates and their consequences for the formation of kin groups. In *Kinship and Behavior in Primates*, ed. B. Chapais and C. M. Berman, 71–108. Oxford: Oxford University Press.

Jerison, H. J. 1973. *Evolution of the Brain and Intelligence*. New York: Academic Press.

Knauft, B. 1991. Violence and sociality in human evolution. *Current Anthropology* 32:391–428.

Leveroni, C., and S. A. Berenbaum. 1998. Early androgen effects on interest in infants: Evidence from children with congenital adrenal hyperplasia. *Developmental Neuropsychology* 14:321–40.

Levi, P. 1979. *If This Is a Man*. New York: Penguin Books.

Lindburg, D. G., ed. 1980. *The Macaques: Studies in Ecology, Behavior, and Evolution*. New York: Van Nostrand Reinhold.

Lindenfors, P. 2004. Females drive primate social evolution. *Proceedings of the Royal Society of London* B 271:S101–3.

———. 2005. Neocortex evolution in primates: The "social brain" is for females. *Biology Letters* 1:407–10.

Lorenz, K. 1966. *On Aggression*. New York: Harcourt Brace Jovanovich.

Machiavelli, N. 1984. *The Prince*. New York: Bantam Books.

Maestripieri, D. 1991. Litter gender composition, food availability, and maternal defense of the young in house mice (*Mus domesticus*). *Behaviour* 116:139–51.

———. 1994. Costs and benefits of maternal aggression in lactating female rhesus macaques. *Primates* 35:443–53.

———. 1995. First steps in the macaque world: Do rhesus mothers encourage their infants' independent locomotion? *Animal Behaviour* 49:1541–49.

———. 1996. Primate cognition and the bared-teeth display: A reevaluation of the concept of formal dominance. *Journal of Comparative Psychology* 110:402–5.

———. 1999. Formal dominance: The emperor's new clothes? *Journal of Comparative Psychology* 113:96–98.

———. 2001a. Intraspecific variability in parenting style: The role of the social environment. *Ethology* 107:237–48.

———. 2001b. Is there mother-infant bonding in primates? *Developmental Review* 21:93–120.

———. 2002. Parent-offspring conflict in primates. *International Journal of Primatology* 23:923–51.

———. 2004. Genetic aspects of mother-offspring conflict in rhesus macaques. *Behavioral Ecology and Sociobiology* 55:381–87.

———. 2005a. Early experience affects the intergenerational transmission of infant abuse in rhesus monkeys. *Proceedings of the National Academy of Sciences USA* 102:9726–29.

———. 2005b. Gestural communication in three species of macaques (*Macaca mulatta, M. nemestrina, M. arctoides*): Use of signals in relation to dominance and social context. *Gesture* 5:57–73.

———. 2005c. Improbable antics: Notes from a gorilla guru. In *King Kong Is Back!* ed. D. Brin, 85–91. Dallas, TX: BenBella Books.

Maestripieri, D., and K. M. Durante. 2004. Infant colic: Re-evaluating the adaptive hypotheses. *Behavioral and Brain Sciences* 27:468–69.

Maestripieri, D., and S. Pelka. 2002. Sex differences in interest in infants across the lifespan: A biological adaptation for parenting? *Human Nature* 13:327–44.

Maestripieri, D., and K. Wallen. 1997. Affiliative and submissive communication in rhesus macaques. *Primates* 38:127–38.

Manson, J. H. 1995. Do female rhesus macaques choose novel males? *American Journal of Primatology* 37:285–96.

Manson, J. H., and S. Perry. 1993. Inbreeding avoidance in rhesus macaques: Whose choice? *American Journal of Physical Anthropology* 90:335–44.

Marais, E. N. 1939. *My Friends the Baboons.* New York: Robert M McBride & Co.

Melnick, D. J., and K. K. Kidd. 1983. The genetic consequences of social group fission in a wild population of rhesus monkeys (*Macaca mulatta*). *Behavioral Ecology and Sociobiology* 12:229–36.

Michael, R. P., E. B. Keverne, and R. W. Bonsall. 1971. Pheromones: Isolation of male sex attractants from a female primate. *Science* 172:964–66.

Michael, R. P., and J. Welegalla. 1968. Ovarian hormones and the sexual behaviour of the female rhesus monkey (*Macaca mulatta*) under laboratory conditions. *Journal of Endocrinology* 41:407–20.

Morgan, C. J. 1978. Bystander intervention: Experimental test of a formal model. *Journal of Personality and Social Psychology* 36:43–55.

Morris, R. 1996. *Partners in Power: The Clintons and Their America.* New York: Henry Holt.

Ordog, T., M. D. Chen, K. T. O'Byrne, J. R. Goldsmith, M. A. Connaughton, J. Hotchkiss, and E. Knobil. 1998. On the mechanism of lactational anovulation in the rhesus monkey. *American Journal of Physiology—Endocrinology and Metabolism* 274:E665–76.

Owren, M. J., D. Rendall, and J. Bachorowski. 2003. Nonlinguistic vocal communication. In *Primate Psychology*, ed. D. Maestripieri, 359–94. Cambridge, MA: Harvard University Press.

Packer, C. 1977. Reciprocal altruism in *Papio anubis*. *Nature* 265:441–43.

Parker, R. 1996. *Mother Love—Mother Hate: The Power of Maternal Ambivalence.* New York: Basic Books.

Pfungst, O. 1911. *Clever Hans (The Horse of Mr. Von Osten): A Contribution to Experimental Animal and Human Psychology.* New York: Henry Holt.

Rawlins, R. G., and M. J. Kessler, eds. 1986a. *The Cayo Santiago Macaques: History, Behavior, and Biology.* Albany: SUNY Press.

———. 1986b. The history of the Cayo Santiago colony. In *The Cayo Santiago Macaques: History, Behavior, and Biology*, ed. R. G. Rawlins and M. J. Kessler, 13–45. Albany: SUNY Press.

Richard, A. F., S. J. Goldstein, and R. E. Dewar. 1989. Weed macaques: The evolutionary implications of macaque feeding ecology. *International Journal of Primatology* 10:569–94.

Rodseth, L., R. W. Wrangham, A. M. Harrigan, and B. B. Smuts. 1991. The human community as a primate society. *Current Anthropology* 32:221–41.

Roney, J. R., K. N. Hanson, K. M. Durante, and D. Maestripieri. 2006. Reading men's faces: Women's mate attractiveness judgments track men's testosterone and interest in infants. *Proceedings of the Royal Society of London* B 273:2169–75.

Sapolsky, R. M. 2005. The influence of social hierarchy on primate health. *Science* 308:648–52.

Savage-Rumbaugh, E. S., and R. Lewin. 1994. *Kanzi: The Ape at the Brink of the Human Mind.* New York: Wiley.

Sawaguchi, T. 1992. The size of the neocortex in relation to ecology and social structure in monkeys and apes. *Folia Primatologica* 58:131–45.

Schino, G., and A. Troisi. 2004. Neonatal abandonment in Japanese macaques. *American Journal of Physical Anthropology* 126:447–52.

Schulman, S. R., and B. Chapais. 1980. Reproductive value and rank relations among macaque sisters. *American Naturalist* 115:580–93.

Sherman, P. W., H. K. Reeve, and D. W. Pfennig. 1997. Recognition systems. In *Behavioural Ecology: An Evolutionary Approach*, 4th ed., ed. J. R. Krebs and N. B. Davies, 69–96. Oxford: Blackwell Scientific.

Simpson, M. J. A., A. E. Simpson, J. Hooley, and M. Zunz. 1981. Infant-related influences on birth intervals in rhesus monkeys. *Nature* 290:49–51.

Soltis, J. 2004. The signal functions of early infant crying. *Behavioral and Brain Sciences* 27:433–59.

Soltis, J., and R. McElreath. 2001. Can females gain extra paternal investment by mating with multiple males? A game theoretic approach. *American Naturalist* 158:519–29.

Southwick, C. H., Y. Zhang, H. Jiang, Z. Liu, and W. Qu. 1996. Population ecology of rhesus macaques in tropical and temperate habitats in China. In *Evolution and Ecology of Macaque Societies*, ed. J. E. Fa and D. G. Lindburg, 95–105. Cambridge: Cambridge University Press.

Thierry, B. 2000. Covariation of conflict management patterns across macaque species. In *Natural Conflict Resolution*, ed. F. Aureli and F. B. M de Waal, 106–28. Berkeley: University of California Press.

Thierry, B., M. Singh, and W. Kaumanns, eds. 2004. *Macaque Societies: A Model for the Study of Social Organization.* Cambridge: Cambridge University Press.

Tomasello, M., and J. Call. 1997. *Primate Cognition.* Oxford: Oxford University Press.

Tomasello, M., M. Carpenter, J. Call, T. Behne, and H. Moll. 2005. Understanding and sharing intentions: The origins of cultural cognition. *Behavioral and Brain Sciences* 28:675–91.

Tooby, J., and L. Cosmides. 1992. The psychological foundations of culture. In *The Adapted Mind: Evolutionary Psychology and the Generation of Culture*, ed. J. H. Barkow, L. Cosmides, and J. Tooby, 19–136. Oxford: Oxford University Press.

Trivers, R. L. 1972. Parental investment and sexual selection. In *Sexual Selection and the Descent of Man*, ed. B. Campbell, 136–79. Chicago: Aldine.

———. 1974. Parent-offspring conflict. *American Zoologist* 14:249–64.

Vahed, K. 1998. The function of nuptial feeding in insects: A review of empirical studies. *Biological Reviews* 73:43–78.

van Hooff, J. A. R. A. M. 1972. A comparative approach to the phylogeny of laughter and smiling. In *Non-Verbal Communication*, ed. R. A. Hinde, 209–41. Cambridge: Cambridge University Press.

van Schaik, C. P. 1989. The ecology of female social relationships amongst female primates. In *Comparative Socioecology: The Behavioural Ecology of Humans and Other Mammals*, ed. V. Standen and R. Foley, 195–218. Oxford: Blackwell.

van Schaik, C. P., R. O. Deaner, and M. Y. Merrill. 1999. The conditions for tool use in primates: Implications for the evolution of material culture. *Journal of Human Evolution* 36:719–41.

van Schaik, C. P., and C. H. Janson, eds. 2000. *Infanticide by Males and Its Implications*. Cambridge: Cambridge University Press.

Vehrencamp, S. L. 1983. A model for the evolution of despotic versus egalitarian societies. *Animal Behaviour* 31:667–82.

vom Saal, F. S. 1985. Time-contingent change in infanticide and parental behavior induced by ejaculation in male mice. *Physiology and Behavior* 34:7–15.

Wallen, K. 1982. Influence of female hormonal state on rhesus sexual behavior varies with space for social interaction. *Science* 217:375–77.

———. 1990. Desire and ability: Hormones and the regulation of female sexual behavior. *Neuroscience and Biobehavioral Reviews* 14:233–41.

———. 1995. The evolution of female sexual desire. In *Sexual Nature, Sexual Culture*, ed. P. Abramson and S. Pinkerton, 57–79. Chicago: University of Chicago Press.

Warren, M. P., and J. Brooks-Gunn. 1989. Delayed menarche in athletes: The role of low energy intake and eating disorders and their relation to bone density. In *Hormones and Sport*, ed. Z. Laron and A. D. Rogol, 41–54. Serono Symposia Publications, vol. 55. New York: Raven Press.

Wasser, S. K. 1983. Reproductive competition and cooperation among female yellow baboons. In *Social Behavior of Female Vertebrates*, ed. S. K. Wasser, 349–90. New York: Academic Press.

Weiner, J. 1995. *The Beak of the Finch*. New York: Vintage.

Westermarck, E. A. 1921. *The History of Human Marriage*. London: Macmillan.

Whitham, J. C., M. S. Gerald, and D. Maestripieri. 2007. Intended receivers and functional significance of grunt and girney vocalizations in free-ranging female rhesus macaques. *Ethology*, in press.

Widdig, A., F. B. Bercovitch, W. J. Streich, U. Sauermann, P. Nurnberg, and M. Krawczak. 2003. A longitudinal analysis of reproductive skew in male rhesus macaques. *Proceedings of the Royal Society of London* B 271:819–26.

Widdig, A., P. Nurnberg, M. Krawczak, W. J. Streich, and F. B. Bercovitch. 2001. Paternal relatedness and age proximity regulate social relationships among adult female rhesus macaques. *Proceedings of the National Academy of Sciences USA* 98:13769–73.

Widdig, A., W. J. Streich, P. Nurnberg, P. J. P. Croucher, F. B. Bercovitch, and

M. Krawczak. 2006. Paternal kin bias in the agonistic interventions of adult female rhesus macaques (*Macaca mulatta*). *Behavioral Ecology and Sociobiology* 61:205–14.

Wrangham, R. W. 1979. On the evolution of ape social systems. *Social Science Information* 18:334–68.

Zehr, J. L., D. Maestripieri, and K. Wallen. 1998. Estrogen increases female sexual initiation independent of male responsiveness in rhesus monkeys. *Hormones and Behavior* 33:95–103.

Zuckerman, S. 1932. *The Social Life of Monkeys and Apes*. London: Kegan Paul.

Acknowledgments

My research and this book could not have been possible without the support of my family. My parents, Elena and Franco, and my sister Daniela shared with me their love for reading and have encouraged me to pursue intellectual activities since I was a child. My wife Kelly supported my work and the writing of this book with great patience and understanding. She also read and edited the entire manuscript. My children Elena and Luca have been a great source of joy, inspiration, and motivation.

I would like to thank the following people for reading some or all of the chapters of this book and providing helpful comments and suggestions: Fausto Cattaneo, Kristina Durante, Melissa Gerald, Bill Hopkins, Joe Manson, Rob Poh, and Wendy Saltzman. Christie Henry, my editor at the University of Chicago Press, encouraged this project from the beginning and provided support at every stage. Working with her has been a great pleasure.

Index